T0123339

Praise for
Helen Scales's

POSEIDON'S STEED

"A compelling book about seahorses that makes the case not only for these odd fish but also for the entire ocean. . . . A nuanced and thoughtful analysis . . . a lucid distillation of a lot of research and careful thought."   —*National Geographic*

"In this fascinating book Helen Scales, an aptly named marine biologist, explains the myth, biology, and ecology of what the Victorians called 'queer fish.'"   —*The Economist*

"Elegant and engaging."   —*Natural History Magazine*

"Scales follows the seahorse in all its manifestations in this delightful book, entrancing readers with tales of plundered treasure, depictions of seahorses in cultures ranging from the Aborigines to the Picts and the Romans, and the history of the ornamental aquarium. . . . This is one charming book about one charming fish."   —*Booklist*

"The author is adept at delineating the seahorse's alleged healing powers, and she offers a fascinating study in the history of aquariums and the pursuit of 'queer fish' . . . a solid case for a rare and wondrous creature."   —*Kirkus Reviews*

"A true natural history book, covering all aspects of the sea-horse's involvement in the world. . . . Scales is a marine biologist, and her fascination with the subject (she learned to scuba dive in order to observe this remarkable creature) shines through."

—*Library Journal*

"This seems to be just about the perfect book: small, delicate, elegant, charming, unusual, fascinating, and uniquely memorable, a classic of its kind. In fact, now I come to think of it, *Poseidon's Steed* is itself a sort of seahorse of the book world."

—Simon Winchester

"This gem of a book has all the charm, passion, and compassion that one could look for in a great, relaxing read. Helen Scales is such a delightful writer and traveling guide that you won't even realize how much you're learning."

—Carl Safina, author of *Song for the Blue Ocean,*
Eye of the Albatross, and *Voyage of the Turtle*

"In this eye-opening book, Helen Scales reveals the heretofore well-kept secrets of what is probably the most fascinating and enigmatic of all the fish in the sea, the tiny, horse-headed, vertical-swimming, tube-mouthed, prehensile-tailed, male brood-pouched hippocampus."   —Richard Ellis, author of *Tuna: A Love Story*

Ria Mishal Cooke

Dr. Helen Scales is a marine biologist and scuba diver who received her Ph.D. from the University of Cambridge. She is a fellow of the Royal Geographical Society and has lived in various countries, conducting research on rare coral reef fish and working for a number of conservation organizations. She is a presenter on BBC radio shows *The Naked Scientists* and *Home Planet,* writes regularly for the *National Geographic* Web site, and has published scientific studies on the trade in endangered wildlife. She currently resides in Cambridge, England.

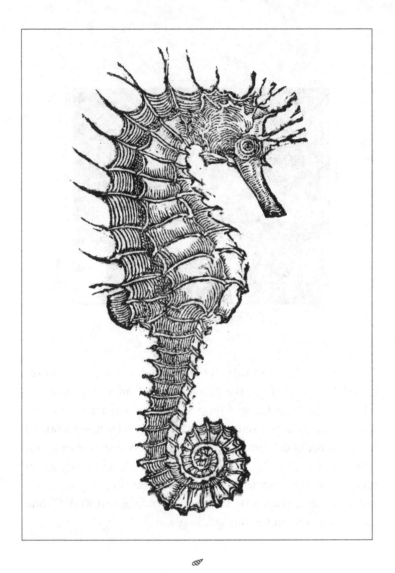

Seahorse woodcut from Guillaume Rondelet's 16th century
Libri de Piscibus Marinis

Credit: Guillaume Rondelet, c. 1555. Reproduced by kind permission of the Syndics of
Cambridge University Library. CCA. 46. 45

POSEIDON'S STEED

The Story of Seahorses, from Myth to Reality

HELEN SCALES, PH.D.

GOTHAM BOOKS

For my Mum and Dad, and for Ivan

GOTHAM BOOKS
Published by Penguin Group (USA) Inc.
375 Hudson Street, New York, New York 10014, U.S.A.

Penguin Group (Canada), 90 Eglinton Avenue East, Suite 700, Toronto, Ontario M4P 2Y3, Canada (a division of Pearson Penguin Canada Inc.) • Penguin Books Ltd, 80 Strand, London WC2R 0RL, England • Penguin Ireland, 25 St Stephen's Green, Dublin 2, Ireland (a division of Penguin Books Ltd) • Penguin Group (Australia), 250 Camberwell Road, Camberwell, Victoria 3124, Australia (a division of Pearson Australia Group Pty Ltd) • Penguin Books India Pvt Ltd, 11 Community Centre, Panchsheel Park, New Delhi–110 017, India • Penguin Group (NZ), 67 Apollo Drive, Rosedale, North Shore 0632, New Zealand (a division of Pearson New Zealand Ltd) • Penguin Books (South Africa) (Pty) Ltd, 24 Sturdee Avenue, Rosebank, Johannesburg 2196, South Africa

Penguin Books Ltd, Registered Offices: 80 Strand, London WC2R 0RL, England

Published by Gotham Books, a member of Penguin Group (USA) Inc.

Previously published as a Gotham Books hardcover edition

First trade paperback printing, November 2010

Gotham Books and the skyscraper logo are trademarks of Penguin Group (USA) Inc.

Copyright © 2009 by Helen Scales
All rights reserved

ISBN: 978-2-592-40581-7
ISBN 978-1-592-40581-7

Set in ITC Galliard   •   Designed by Elke Sigal

Without limiting the rights under copyright reserved above, no part of this publication may be reproduced, stored in or introduced into a retrieval system, or transmitted, in any form, or by any means (electronic, mechanical, photocopying, recording, or otherwise), without the prior written permission of both the copyright owner and the above publisher of this book.

The scanning, uploading, and distribution of this book via the Internet or via any other means without the permission of the publisher is illegal and punishable by law. Please purchase only authorized electronic editions, and do not participate in or encourage electronic piracy of copyrighted materials. Your support of the author's rights is appreciated.

While the author has made every effort to provide accurate telephone numbers and Internet addresses at the time of publication, neither the publisher nor the author assumes any responsibility for errors, or for changes that occur after publication. Further, the publisher does not have any control over and does not assume any responsibility for author or third-party Web sites or their content.

146119709

CONTENTS

✥

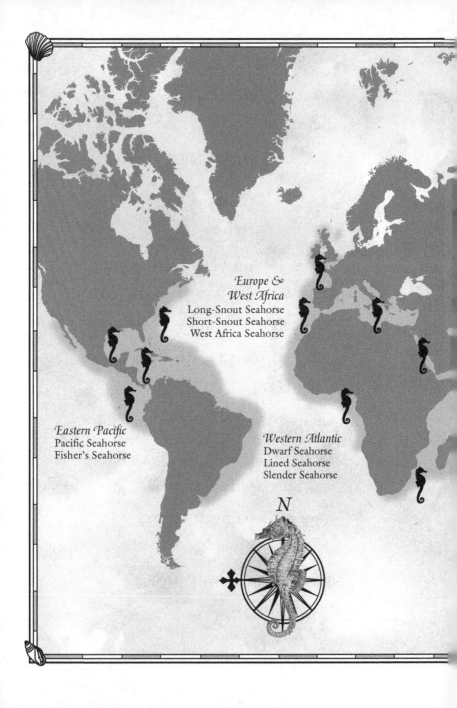

Europe &
West Africa
Long-Snout Seahorse
Short-Snout Seahorse
West Africa Seahorse

Eastern Pacific
Pacific Seahorse
Fisher's Seahorse

Western Atlantic
Dwarf Seahorse
Lined Seahorse
Slender Seahorse

N

Where to Spot a Seahorse
A Map of Seahorse Habitats

Japan
Lemur-Tail Seahorse
Shiho's Seahorse
Crowned Seahorse

Southeast Asia
& West Pacific
Bargibant's Seahorse
Denise's Pygmy Seahorse
Zebra-Snout Seahorse
Tiger Tail Seahorse
Thorny Seahorse
Kellogg's Seahorse
Yellow Seahorse
Hedgehog Seahorse
Severn's Pygmy Seahorse
Pontoh's Pygmy Seahorse
Satomi's Pygmy Seahorse
Walea Pygmy Seahorse

Indian Ocean,
Red Sea, &
Persian Gulf
Réunion Seahorse
Giraffe Seahorse
Knysna Seahorse
Knobby Seahorse
Sea Pony
Jayakar's Seahorse
Lichtenstein Seahorse
Soft Coral Seahorse

Australia
& New Zealand
Big-Belly Seahorse
Narrow-Belly Seahorse
West Australian Seahorse
Knobby Seahorse
Bullneck Seahorse
White's Seahorse
Zebra Seahorse
Coleman's Pygmy Seahorse

PRELUDE

And the Sea-horse, though the ocean
Yield him no domestic cave,
Slumbers without sense of motion,
Couched upon the rocking wave

William Wordsworth, "Song for the Wandering Jew," 1827

*P*eer at a seahorse, briefly hold one up to the light, and you will see a most unlikely creature, something that you would hardly believe was real were it not lying there in the palm of your hand, squirming for water. Should we presume these odd-looking creatures were designed by a mischievous god who had some time on her hands? Rummaging through a box labeled "spare parts" she finds a horse's head and, feeling a desire for experimentation, places it on top of the pouched torso of a kangaroo. This playful god adds a pair of swiveling chameleon eyes and the prehensile tail of a tree-dwelling monkey for embellishment—then she stands back to admire her work. Not bad, but how about a suit of magical color-changing armor, a perfect fit, and a crown borrowed from a fairy princess, shaped as intricately and uniquely as a human fingerprint? Shrink it all down to the size of a chess piece and the new creature is complete.

No matter how tempting such a strange tale of creation may be, seahorses are real creatures, a product of natural selection. They inhabit a wide stretch of the oceans and are not, as we might suppose, restricted to warm azure waters that lap on equatorial shores. If you stand with your toes dabbling in shallow sea almost anywhere in the world—except for the iciest spots where your feet would freeze—there is a chance you might see a seahorse. Not a very great chance, admittedly, but a chance nonetheless.

What is it that makes seahorses seem so special, like miniature dragons of the sea? Where does their peculiar appearance come from, why do they look like nothing else on earth? What goes on during a day in the life of a seahorse? And how did they evolve to be the only species in the world in which males give birth?

As a marine biologist, I tinker with these sorts of questions, the questions that fascinate me and occasionally keep me awake at night and certainly move me to do the things I do.

On an inauspicious day hemmed in beneath low gray skies I stood, aged sixteen, on the shores of a flooded gravel pit in the British Midlands and decided what I wanted to "be." I had just returned from my first dive in open water—my first underwater foray away from the sterile chlorination of the swimming pool, a chance to put months of training into practice. The greatest shock on that dive had been the first thirty seconds as the terrifying cold crept in through the cuffs of my inelegant, thick "semi-dry" suit—there was obviously nothing dry about mine. A single fuzzy notion stammered around my gradually freezing brain: "Do people really do this for fun?" It wasn't long before my inexperienced feet drifted too close to the squidgy pit bottom, stirring up a turbid brown haze and adding vision to my list of rapidly vanishing senses. Only then was it obvious why my instructor had insisted on tethering me to him with a plastic cord, like a dog on a leash. I couldn't see my hand in front of my face, let alone another diver a few feet away. After relentless, perishing minutes of wishing this whole dreadful business could be over and forgotten, something happened that instantly made it all worthwhile. Through a parting in the murky water I saw a fish. It wasn't a particularly pretty fish, just small and silvery and, frankly, quite normal-looking. But even so, it was a wild fish and I was sharing its three-dimensional aquatic realm. Despite the rowdy bubbling of my human exhalations and my amplified clumsiness in the numbing cold, I caught a brief, glorious glimpse of what it is like to be a fish. All it took was this single unremarkable fish to hook me in, utterly and irretrievably, and transform me into a marine biologist in the making.[1]

Picture, then, what went through my mind eighteen months later when, as a fully qualified diver, I leapt into tropical waters for the first time. The sea off Belize was warm like a bath and so clear I could see the coral reef spread out around me in every direction for thirty meters or more. With childlike glee, I frolicked through a vibrant, kaleidoscopic sweet shop. Electric blue fish darted between glowing lime-green towers of sponges. Shoals of snappers, with banana yellow stripes, spilled over the reef edge into the bottomless cobalt depths. "The color blue has never held such meaning to me," a line in my dive logbook reads. For eight blissful weeks I camped on a desert island, learned how to identify corals and fish, and spent hours exploring and studying a submerged wonderland. I met my first sharks and sea turtles, swam with stingrays that were bigger than me and I watched, speechless, as flying fish leapt from their watery realm and skittered across the tops of the waves. That trip confirmed my instinct on seeing that first fish back in the British gravel pit:

Marine biology was for me. It was also in Belize that I first became preoccupied with the idea of seeing a particular group of aquatic animals. Of all the hundreds of thousands of fish that live in the seas, I quickly realized it was the seahorses I wanted to see the most. Something about them felt subtly irresistible to me, something to do with their perplexing appearance tangled up in my longing to understand their obscure lives. From then on, no matter where I went diving or what I was supposed to be doing while I was down there, I began to keep an eye out for the silhouette of a down-turned snout or the twitch of a chameleon-like eye. But I was going to have to wait a long time before my first encounter with a wild seahorse.

In the years that followed my induction into the tropics, through countless semesters of university study, summer vacation work, and research jobs around the world, my head filled up with a medley of marine encounters, each one deepening my devotion to the oceans. After finishing high school, I delayed my entry to university for a year, grabbing the opportunities that a gap year would offer me to explore my beloved marine domain. While planning my travels, I watched a TV documentary about whale sharks and was instantly determined to see one for myself. I set off for Australia's remote west coast, offering my services as a volunteer with a whale shark research group, and in a rendezvous I will never forget, found myself swimming alongside the undisputed emperor of the fish world. "Over there," the boat captain shouted above the puttering of the engine, waving his arm across the choppy Indian Ocean waves. I couldn't see a thing as I jumped in the water and kicked as hard as I could in the vague direction of his gesture. Whale sharks are big, monstrously so, and nothing can prepare you for seeing one up close; being in the water next to one creates a diminutive feeling like no other. It steamed toward me like a train. I wasn't scared it would run me down or eat me; they are, after all, harmless plankton feeders. I was simply staggered that it was getting closer and closer and still there was no end in sight. As its wide square jaw approached, it peered at me with a deep-set piggy eye. Its inky-blue skin, stippled in fist-size white spots, slid past me like an unending conveyor belt, until eventually its scythe-like tail fin planed into view; the tail alone was taller than I was, even with my dive fins on. I felt honored and belittled by the brief moments we spent together, before it powered off, too fast for me to keep up.

Another set of my treasured snapshots come from time I spent on Layang Layang, a tiny island in the South China Sea, hundreds of miles off the coast of Borneo. For several months while I studied for my Ph.D., my daily commute to work involved a half-hour boat ride across a pale emerald lagoon accompanied by a pod of spinner dolphins, the same ones each time, marked with individual patterns of spots and scratches. They would frolic in the boat's bow wave and look up at me as I leaned over the side to listen to their trills and whistles. The purpose of my trip was to study one of the largest remaining populations of a highly endangered coral reef fish called the Napoleon wrasse, known also—for obvious reasons when you see one—as the humphead wrasse. These are fish that live for decades—starting off as females, changing sex later in life—and grow so large they wouldn't fit in a bathtub. They are hunted to supply a growing fashion for "live" seafood restaurants across Asia. Tableside aquariums clamor with fish, air-freighted in from reefs around the world, where they wait to be selected by rich diners before being whisked off to the kitchen and presented to the banqueting table within minutes of dying. The tender flesh of Napoleon wrasse is at the top of the price list, as well as plates of their rubbery lips—the ultimate delicacy and status symbol to impress dinner guests. I have never eaten Napoleon wrasse, no one has ever offered, but I have hovered in a thousand feet of open sea while all around me a frenzied forty-strong maelstrom of them were busy making babies. I visited one of only a handful of locations in the world where anyone has witnessed the spectacular event of Napoleon wrasse spawning. They behave more like amorous rain forest birds or lions in the African savannah than fish in the ocean, with strict dominance hierarchies, fiercely defended territories, and elaborate courting rituals. Once or twice, the dominant male charged at me, perhaps thinking I was an intruding male competitor. In reality, all I was trying to do was photograph him and his harem and document the scribbles and spots that are unique to each fish's face. Using my gallery of portraits, I discovered that the females can't get enough of the male Napoleon wrasse's charm and they come back day after day for more tumultuous spawning action.[2]

My time as a marine biologist has not all been about diving in crystal waters and gazing at iridescent fishes. During my undergraduate degree at Cambridge University, I wrote a dissertation on coral taxonomy, the painstaking science of identifying species. Each afternoon for five months, I locked myself away in fluorescent solitude in a Cambridge basement and, armed with a microscope, I worked my way through a teetering mountain of dead corals. It was about as far from the sea as I could get. Then, after graduating, I escaped once more to the tropics and took a job with a conservation organization in Malaysia.[3] For a year I earned forty Malaysian ringgit a day (about ten U.S. dollars) in return for romping around an archipelago of a hundred tropical islands to catalogue all the marine creatures that lived there. A dream job—surely—but not when I arrived to discover that the unchartered reefs I was tasked to explore were shrouded within soupy, sediment-laden waters. The sea may have been a sultry eighty-four degrees Fahrenheit (twenty-nine degrees Celsius), but once again, like that first British dive, I couldn't see far beyond the glass in my dive mask.

My research and travels have also brought with them some unexpected experiences, like when a Triad gang leader in Malaysia introduced me to his network of fish traders over dinner and served us a plate of chewy sea cucumbers (the cat under the table was the lucky recipient of most of my meal that night). Or inventing a way to rescue a sunken outboard engine from the bottom of the sea. I take pride in my tough stomach, but there have been times when heaving seas have taken their toll. Occasionally, as when I was bobbing around California's Monterey Bay in a small sport fishing boat trying to tag sharks or when the boat I was on broke down miles from shore in a turbulent patch of the South Pacific, it all gets to be too much and I long for land.

During more than a decade of diving and studying the seas, the astonishing wildlife and adventures haven't been the only things driving me on. I have also watched the fraying edges of ocean ecosystems unraveling before my eyes. Each time I see changes taking place firsthand, I feel like the ocean flows deeper into my veins. The first time I bore witness to the shifting oceans was on my first trip to tropical shores. Toward the end of my visit to Belize, a front of warm water arrived, a symptom, some say, of climate change. Within days, parts of the once brightly colored reefs surrounding Belize transformed into an eerie pale landscape as the colored microscopic algae living inside the corals became heat stressed and abandoned their hosts. That year, 1995, was the first time people had seen the reefs of Belize suffering from mass coral bleaching. It has happened several times since then, the worst in 1998, when nearly half of Belize's corals died.[4]

Once, while diving in Malaysia, I learned what it feels like to

be the target of an ecologically disastrous but increasingly common method of fishing known as a "fish bomb." It was the only
time I have ever been genuinely terrified underwater. The pressure wave from the makeshift explosive tore into me, battering
my eardrums and thundering through my insides. All I could
think of was to escape from the water before another bomb
went off that might deafen or even kill me. Arriving back at the
surface, I was staggered to see how far the bomb's reverberations had traveled, all the way from a fishing boat that idled almost out of sight, hundreds of meters away.

Throughout all the privileged time I have spent underwater,
after all the thousands of species I've seen, the seahorses have
hovered, stubbornly, out of sight. Many times I have watched
them in captivity, with my nose pressed up to the glass tanks
of public aquariums, mesmerized by their perfect minute features and cautious, graceful movements. But for many years I still
hadn't seen one in the open sea. I dreamed of what it would be
like to see one, to proudly write "seahorse" in my dive logbook,
to peek at one as it went about its daily life in its natural habitat
and not have to make do with staring at one that was confined
to a man-made tank.

And it wasn't as if I hadn't tried to find a wild seahorse. On
several occasions I set out single-mindedly to find a seahorse
and several times I nearly succeeded. In the South China Sea,
after long days spent photographing Napoleon wrasse, I would
go night diving inside the atoll's shallow lagoon and scour the
lush seagrass meadows, hunting in vain for the glint in a pair of
seahorse eyes shining in my flashlight beam. In Mauritius, a local dive master took me to a spot inside the pellucid turquoise
lagoon that cocoons the Indian Ocean island and where, he
promised me, he always saw seahorses, or "eeppocamps" in his

native French. As soon as we arrived at the sandy seabed five meters below, and before I could refuse, he thrust into my hands a long black bootlace that turned out to be a bend-stick pipefish. It was almost a seahorse, with an angled head and not-so-straight tail, but not quite.

During a trip to Borneo, a local friend of mine, Huan, heard stories of fishermen catching seahorses in shrimp nets from nearby seagrass beds, so we grabbed our snorkeling gear and went to see for ourselves. After hours of fruitless searching, Huan excitedly shrieked from a way off, "Helen, Helen! Come quick!" My heart raced as I hurried over, thinking that my time had finally come, then it sank like a dive weight dropped over the side of a boat when I saw what Huan was pointing at: a meter-long sea cucumber called a synaptid, draped in great loops on the seafloor like a giant marine caterpillar. "See how it's feeding with those feathery appendages," ever-enthusiastic Huan said, grinning and unaware at how he had raised—and then dashed—my hopes of seeing an elusive seahorse that after-noon. (To this day I still feel slightly glum wherever I see a syn-aptid lying hopelessly on the seafloor, defecating long sausages of compacted sand.)

Some time later, I did see a Bornean seahorse, but I decided that it didn't really count. I was conducting surveys for a pro-posed shrimp farm in shallow bays at the northernmost tip of the island as it points toward the Philippines. The boat driver had found a seahorse and brought it to me in a plastic bucket. I released it back into the sea and felt guilty for days, thinking we had separated it from a faithful partner that, otherwise, it might have spent the rest of its life with, as many monogamous sea-horses do.

It wasn't until a recent trip to Vietnam, fourteen years af-ter my first open-water dive, that I finally met a seahorse in its ocean home. During a window in my busy work schedule, I visited a dive site I was told had a well-stocked paddock teeming with grazing seahorses. The site was on the underwater flanks of a ragged island called Mama Hanh, supposedly named after a famous local lady because of its sensuous, feminine profile—I wasn't so sure. I hopped eagerly into my dive gear and, along with my dive buddy, slipped beneath the water surface, smooth like a swimming pool before anyone has jumped in. I drifted down and was immediately engulfed by the familiar and blissful sensation of abandoning gravity, as a blue ceiling closed over my head and trails of silver baubles streamed out behind me, rising up into the distance. As patches of coral appeared below my fins, I felt a twinge of expectation. Perhaps this time I would find one?

For a while my hopes were high. I got my "eye in," spying minute camouflaged critters; I counted dozens of sea slugs like candy-striped nail clippings fallen down between narrow coral crevices and inch-long, buff-colored gobies resting on their front fins like little legs with pairs of "I'm-watching-you-up-there, really-I-am" eyespots drawn on the tops of their heads. I even saw one of the seahorses' extremely rare relatives, the ornate ghost pipefish,[5] pretending it was not there amid the curling tendrils of a feather star. But as the dive computer on my wrist ticked away, I resigned myself to another near miss. Forty minutes had passed and all the divers would soon be expected back on the boat. Then suddenly, the metallic clang of dive knife on tank made me flinch; my dive buddy was trying to catch my attention. He was hovering a few meters away above a small cluster of dead-looking coral, pointing triumphantly. There it was—pumpkin orange, covered in short prickles and with a pair of white saddles painted across its back. It was not especially big, only about as tall as the space between my index finger and thumb curved in a C-shape, and it was snoozing quietly on the seafloor, its tail wrapped neatly around a broken finger of coral. The funny thing was, every time I had played out this moment in my mind and imagined what it would be like to find my first wild seahorse, I had got it wrong. There was no doubting it was as beautiful as the seahorse in my daydreams, but when I saw it, I didn't shriek in muffled delight or dance around in circles of excited bubbles. Instead, I was clutched by a deep serenity and the overwhelming urge to lie belly-down on the sand, my chin resting in my hands, and simply watch it: forever, if I could.

What was so special about that encounter with my first seahorse? Why are seahorses different from so many other creatures I have gazed upon during hours and days underwater? It surely has something to do with their unconventional beauty, their unique combination of features, a demure down-turned snout and tightly curled tail, as sensitive and nimble as an elephant's trunk. Was it the anticipation of seeing one and the satisfaction, after such a long wait, of finally spying one of these wily masters of disguise that can match themselves so expertly to their surroundings? Sharks are breathtaking in their nonchalant efficiency as they slice through the water like sleek torpedoes. Reef fish are instantly gratifying, colorful and brazen, flitting around like butterflies of the sea. But seahorses hold a secret intimacy, a special reward for the keen-eyed. And perhaps deep down I held on to a childhood suspicion, an irrational part of me that didn't quite believe seahorses really do inhabit the oceans. Seeing one felt like glimpsing a unicorn trotting through my garden.

Whatever it is about them, I am certainly not alone in my fascination with seahorses. For thousands of years, they have quietly tugged at mankind's imagination, and tenaciously captured our attention. Our ancestors puzzled over their odd appearance, wondered where they came from and yearned to make sense of what they are. On seeing them for the first time in the New World in 1672, naturalist John Josselyn thought that seahorses were "so like the picture of Saint George his dragon, as possible can be, except his legges and wings."[6]

Over the centuries, seahorses have meant many different things to many different people. They have been a source of artistic and mythical inspiration: a template for fabulous beasts, the basis for great works of art, elegant jewelry, literature, and poetry. They have been talismans to ward off evil spirits, a scientific conundrum waiting to be unpicked, or merely a thing of beauty to be gazed upon. There are people today who think seahorses are worth more than their weight in gold and many who suspect they can't be real creatures at all, and are in fact made-up fairy tales.

As you read this, you have your own pair of private seahorses sparking signals inside your gray matter, responsible for your short-term memories and spatial navigation. A sixteenth-century Italian anatomist, Guilio Arantius, dissected the human brain and found in each hemisphere a region he thought looked like the curvaceous form of a seahorse. He named each nodule "hippocampus," an ancient term for seahorses. Scientists recently discovered that some of us have bigger neural seahorses than others; apparently the paired hippocampi are especially large inside the brains of London taxi drivers. This part of the cabbies' brains appears to play an important role in forming complex mental maps of Central London's maze of streets. To pass a test called "The Knowledge" and earn their all-London license, drivers must memorize the layout of twenty-five thousand streets within a six-mile radius of Charing Cross train station, including the location of every pub, hospital, hotel, police station, and restaurant. It isn't just that people with larger hippocampi are better at remembering routes and thus are more likely to become taxi drivers, but the longer someone has driven taxis, the bigger his or her hippocampi tend to be.[7] It seems our inner seahorses can be trained to deal with the task of helping us find our way around.

Meeting all of the seahorse species and exploring the myriad ways they have been brought into our human lives involves a journey around the world, to the reefs of Indonesia, to the back streets of Hong Kong, to a courtroom in Turkey and fishing boats in Vietnam. We will also need to go back in time, to watch our ancient ancestors carving pictures in stone, to hear them tell tales of sea monsters that have the head of a horse and the tail of a fish and to share a Victorian obsession with nature.

Seahorses are icons of the sea. They are one of the tiniest and most timid oceanic inhabitants, spending much of their lives hidden away, and yet every one of us knows of them—even if we might not quite believe they are real—and we smile when we recognize their eccentric shape. They remind us that we rely on the seas not only to fill our dinner plates but also to feed our imaginations. When we steal them away from their watery homes to fill our stories, dreams, aquariums, and medicine cabinets, seahorses bring with them a message from beneath a rumpled blanket of blue waves. It is a message of how a fragile world of incomparable beauty and eternal intrigue is being slowly unpicked by mankind.

Chapter One

A MIGHTY STEED

*I*nside, the tomb was silent, cool and still. Beneath a thousand-year darkness hid a vast treasure trove, an unimaginable Aladdin's cave. Outside, the hot desert lay stretched out like an unending yellow carpet with eye-stinging sand that danced through the midday wind. It is debatable how men from a nearby village in the Uşak province of Turkey came to discover this buried hoard, but it was Osman the blacksmith who was chosen to detonate the stick of dynamite and be the first one to climb into the tomb. Later he would say that he had been led into grave misfortune and struck, perhaps, by the curse of the tomb. But he was clearly in the habit of taking chances, since that day was the sixth day of the sixth month of 1966—far too many sixes for the superstitious.[1]

Some say the treasure had belonged to King Croesus of ancient Lydia, a man whose wealth was so legendary that we still say

∽

Above: Drawing of hippocamps carved into a Minoan seal

Credit: JHS Evans, 1897

someone is as "rich as Croesus." For several centuries around the
first millennium BC, the kingdom of Lydia flourished amid a jos-
tling of invasions and conquests that spread across the western
flanks of Asia Minor in what is now Turkey. Sixty miles east of
the coastal city of Izmir lie the remains of Sardis, the ancient
capital of Lydia. What little is left of it hints at the existence of
a splendid and prosperous society. The streets were lined with
houses and shops, there was a grand bathhouse and an ornate
synagogue. Many think coins were invented in Sardis around
twenty-seven hundred years ago.[2] The source of Lydia's great
wealth was most probably the gold-laden waters of the river Pactolus, on whose banks the foundations of a gold refinery have
recently been uncovered.[3] Legend has it that the gold was left behind when King Midas, fed up with turning all he touched into
gold, leapt into the river to wash away his despised Midas touch.

History leaves no clear record of what became of King Croesus. Several of his predecessors, including his father, Alyattes, are
thought to be buried in a collection of vast tumuli at a royal cem-
etery called Bin Tepe, an invasion of giant molehills scattered
across the flat Hermus River valley. Known as the "pyramids of
Anatolia," these conical burial mounds were built for the nobil-
ity and rich citizens of Lydia. The largest one measures over a
kilometer around its base and is more than twice as tall as the
Statue of Liberty at its peak. Most of the Lydian tumuli were
rediscovered and plundered for their valuable contents long ago.
But for some unknown reason, the tomb near Uşak remained
untouched and in obscurity for millennia until the fateful day in
1966 when it was broken into by a group of Turkish villagers.

The tranquility of the tomb ended in a blast of noise and
debris. Shafts of hot light rushed in; fresh air mingled with the
ancient atmosphere. The other villagers climbed in after Osman, their shrieks and hollers of delight echoing off the splen-

did tomb walls that were decorated in murals of Lydian men and women, as they discovered sumptuous heaps of silver and gold, platters, statues, goblets, and jewelry.[4] Among the piles of treasure was a tiny, inch-long gold brooch sculpted in the shape of a mythical creature, a winged seahorse, with the head and front legs of a horse, the scaly tail of a fish, and a pair of giant feathered wings. When the tomb raiders first saw the seahorse brooch, one of them might have been tempted for a moment to pocket the small, delicate trinket, to take it home as a gift for his wife so she could gaze at it and trace her finger along the intricate knot work of nine dangling golden braids, each one ending in a fading glass bead. But money was a greater temptation and the brooch was carried outside, along with all the other treasures, which glittered once again in bright daylight. After two-and-a-half thousand years of lying in peaceful darkness, the golden winged seahorse reemerged into a very different world where it would be hurled into the center of an international scandal that reverberated around the world for decades to come.

For a long time, mankind has been fascinated by seahorses. Our ancestors first brought seahorses into the human world as great works of art, as paintings, carvings, and jewelry, attempting to capture the unusual and elegant beauty of these mythical and magical beasts. Seahorse artifacts have lain buried and hidden around the world for hundreds and even thousands of years. One by one they have been uncovered by archaeologists or tomb raiders or farmers plowing their fields, until today we can build a picture of the enduring obsession humans have had with these strange creatures that look like nothing else on earth.

The very earliest signs of our seahorse imaginings appeared as long as six thousand years ago in Australia. When Europeans arrived to colonize the southern continent in the eighteenth century, they encountered hundreds of Aboriginal tribes who had lived in Australia for thousands of years and who believed in an assorted collection of Great Spirit ancestors. These spirits created the world and everything in it; their stories were told in myth, song, and dance; and their images adorned cave walls and rock faces. In an attempt to classify and distill the complex Aboriginal folklore, western anthropologists coined the name "Dreamtime" and began recording and studying the varied beliefs, rituals, and artwork of tribes across Australia. One of the most greatly revered ancestral spirits of the Dreamtime, common to many distinct tribes, was the Rainbow Serpent. Usually portrayed as female, the serpent has slithered through oral histories, ceremonies, and cave art for thousands of years up to the present day, where she symbolizes creative and destructive powers of nature. As Dreamtime legend has it, the Rainbow Serpent emerged from the earth long ago and wandered about, sculpting the landscape into mountains, rivers, and gorges. She now lives on in the present and future of the Dreamtime, inhabiting deep water holes marked by water lilies; she can be seen reaching into the sky as rainbows; she triggers the onset of the wet season and punishes wrongdoers by unleashing natural disasters.[5]

The way Rainbow Serpents are depicted in art varies across Australia from tribe to tribe and has also changed over time. Modern-day serpents painted on strips of string bark are quite obviously composite creatures, assembled from a host of different animals and occasionally plants: the snarling teeth of crocodiles, a pair of perky kangaroo ears, the stalks and leaves of water lilies, and even the horns of buffalo, a species introduced to Australia only in the nineteenth century. The oldest known Rainbow Serpents are from Arnhem Land, in the northwest corner of Australia, in and around the famous Kakadu National Park. Drawn between four and six thousand years ago, the older Rainbow Serpents tend to have fewer multispecies adornments than more recent depictions and instead take on the sinuous form of a snake, perhaps imitating the water pythons that live in this part of the country. Some of these ancient serpents have, however, borrowed distinctive features from another creature. With the serpents' curved, ridged bodies, tubular snouts, and angled heads tucked down toward the chest, there is no mistaking the hint of a seahorse. A few even have a pregnant bulge. Archaeologists have pointed out that many Rainbow Serpent paintings look a lot like a close relative of the seahorses called the ribboned pipefish. These seahorse cousins sprout weedy protrusions along their narrow bodies and live in shallow warm waters along the northwest coast of Australia. Further hints that these serpents were inspired by the sea come from various other objects drawn alongside some early Rainbow Serpents. Wobbly combs could represent sea cucumbers lying on the seafloor, wavy lines could be seaweed wafting in the current, and circles struck through with crosses perhaps give the impression of sea urchins.

The resemblance of early Rainbow Serpents to seahorses could simply be a coincidence, but the timing of their appearance in ancient Australia also lends a persuasive argument to the theories of a watery origin. Archaeological evidence suggests that Aborigines have occupied Australia for at least forty thousand years. This means they would have lived through a time of great change around ten thousand years ago when the planet emerged blinking from its last ice age; ice sheets melted and poured into the oceans, pushing up sea levels and causing shorelines to race inland. Old landscapes were drowned and new ones created. Treacherous, swelling seas and stormy weather would have forced snakes from waterlogged soils, thrown rainbows into the skies, and may even have cast unusual-looking sea creatures up on the shore, including indigenous seahorses and pipefish. It was only at the end of this turbulent time, around six thousand years ago, that pictures of Rainbow Serpents began to emerge on cave walls, leading some archaeologists to associate various aspects of the changing climate with the nature of the ancestral spirit that was born. It could be that the Rainbow Serpent was created as an emblem of the disruption caused by the dying ice age, or as a symbol used by the Aborigines to help make sense of the new landscape and the creatures that were found there.[6]

After Australian Aborigines first painted Rainbow Serpents, the next time seahorses appeared in ancient artwork was on the other side of the world around five thousand years ago. The Greek island of Crete in the Mediterranean Sea was once home to an ancient civilization of Bronze Age people known as the Minoans. They lived in a prosperous agricultural society, cultivating olives and grapes and rearing sheep for their wool. They practiced metalworking and made elegant seals and jewels, among them a three-sided prism embossed with two lifelike seahorses lying head to tail. Many archaeologists believe these were meant to represent the real seahorses that still live in the Mediterranean and would have occasionally washed up on Minoan shores.

Another seahorselike creature that emerged across Europe at around the same time was less naturalistic, taking on a mythological persona with the head, torso, and forelegs of a horse and the tail end of a fish. This was a beast that came to be known as hippocampus, from the ancient Greek words *hippo* meaning "horse" and *kampos* meaning "monster" (or possibly "caterpillar"), a name that real seahorses would later be given. The oldest known hippocampi were carved into stone stamps known as island gems, used across ancient Europe and Asia to emboss intricate pictures in wax and clay, denoting a legal or personal seal. Exactly where the idea for the hippocampus came from remains something of a mystery. It is often assumed that they made their way across from the East like many other marine-themed beasts; mermaids and mermen, humans with fish tails, have indisputable Asian origins, as do the marine dragon Ketos and the fish god Triton. But, so far, an Asian hippocampus has never been found. Wherever they came from, the idea of the half horse, half fish caught on and swept through various ancient European cultures.[7] The Phoenicians were a seafaring race who lived around four thousand years ago along a narrow coastal strip, now part of modern-day Syria and Lebanon, a historic channel of communication between Africa and Asia. They were often painted on tomb walls and coffins alongside other sea creatures that escorted the dead on their voyage across the sea into the afterlife. Before they were taken over by the expanding Roman Empire, the Etruscans from northern Italy embellished their burial chambers in similar maritime friezes. There is even a single appearance of a hippocampus in Egyptian art. In a small museum in the English city of Gloucester lies a mummy's coffin painted with the kneeling figure of the goddess Isis, who looks down on a small seahorse. Its head resembles a child's hobbyhorse, with big round eyes, spots on its muzzle, a long fuzzy mane, and a smiling open mouth, its tail tapering with a lifelike curl.[8]

Perhaps the best-known hippocampi are the ones that feature as popular characters galloping through Greek mythology. Reminiscent of the Australian Aboriginal folklore, the Greek myths tell stories of a pantheon of powerful gods whose adventures explained the origins of the universe and provided metaphors and morals to help people make sense of the world around them. Initially, the stories of ancient Greece were passed on by word of mouth, no doubt elaborated as they descended through the generations, until poets and dramatists began to write down their own versions in the seventh century BC, at around the same time that the Lydians were inventing coins. These stories went on to become so influential, they resonated throughout the culture, art, and literature of western civilization.

Hippocampus was never given a main part in any of the Greek myths, but it would often be seen trotting past in the background of vases and murals, ridden by enchanting sea goddesses known as the Nereids. These captivating, shape-changing nymphs were divine spirits that embodied the fascinating lure of the mysterious deep ocean and who liked nothing more than to spend their time dancing and singing. But hippocampus' most famous master was Poseidon, the Greek god of the sea.

After overthrowing their parents in an ancient Titanic coup, Poseidon and his brothers, Zeus and Hades, drew lots to divide the world among them: The skies were for Zeus, the underworld for Hades, and Poseidon was put in charge of the oceans. He was usually portrayed as a fearsome bearded man who flaunted a trident and lived beneath the waves in a sparkling palace of corals and jewels with his wife, one of the Nereids, Amphitrite. Together they raced around the oceans on his golden chariot, pulled by a quartet of giant hippocampi.[9]

Poseidon not only held dominion over the seas, but with his torrid and tumultuous nature, he often behaved like them, too. Many times he used the seas to vent his unrelenting and unforgiving anger on his enemies, as we see in Homer's epic poems *The Iliad* and *The Odyssey*.

In *The Iliad*, Laomedon, king of Troy, provokes Poseidon's outrage by duping him into building the walls of the city without paying him, and even threatens to enslave the great sea god and cut off his ears. In retaliation, Poseidon sets loose a savage sea monster on the city and, throughout the rest of *The Iliad*, does all he can to help the Greeks in their war against the Trojans. But his siding with the Greeks comes to an end in *The Odyssey* when the Greek hero Odysseus also makes the mistake of crossing the great monarch of the sea.

Following Odysseus' ingenious trick with the wooden horse that ended the Trojan War, the opening passages of *The Odyssey* describe the victorious Greek fleet returning home to Ithaca. But their journey turns out to be neither speedy nor tranquil as the sailors stumble into a string of perilous adventures. Poseidon gets involved when Odysseus picks a fight with one of his numerous offspring, a monstrous, one-eyed cyclops called Polyphemus. Stabbing the cyclops in his one and only eye with a red-hot poker earned Odysseus confinement on a remote desert island, unable to cross the wild seas that were controlled by the enraged Poseidon. The prisoner finally escaped after seven years with the help of Poseidon's brother Zeus when the vengeful sea god was away at a banquet. Realizing what was going on, Poseidon galloped on hippocampi through the oceans and found Odysseus clinging to a wooden raft. Waving his trident toward the heavens, Poseidon "stirred up the sea; he roused the storm-blasts of every kind of wind, and enveloped the land and sea alike in a cloud." The hero of the story is lucky to survive this mighty tempest, kept afloat on a shawl thrown to him by a compassionate sea goddess.

No one knows for sure how literally the ancient Greeks took their mythology, but sailors and fishermen made sure not to incur the sea god's wrath; they built temples to worship him, drowned sacrificial horses in his name, and implored him for safe passage across the seas. Fishermen believed the real sea-horses they occasionally found tangled in their nets were the tiny offspring of Poseidon's mighty steeds.

Poseidon and the other Greek gods did not disappear with the demise of ancient Greece. After the Romans declared victory at the Battle of Corinth in 146 BC, they absorbed many aspects of Greek culture and art, including much of the pantheon of Greek gods. The only thing that really changed were the names of the gods, and Poseidon lived on as the Roman god Neptune, keeping his long beard, his trusty trident, his twinkling submarine palace, and of course his faithful steed.[10] The Roman poet Virgil wrote about Neptune and hippocampus in his epic poem *Aeneid*:

> *Where'er he guides*
> *His finny courses, and in triumph rides,*
> *The waves unruffle and the sea subsides.*

(Virgil, Aeneid I)

Another Roman poet, Statius, wrote about the sea god and his oceangoing journeys in his unfinished poem, *Achilleid*, about the Greek warrior Achilles:

> *Shaking his trident, urges on his steeds,*
> *Who with two feet beat from their brawny breasts*
> *The foaming billows; but their hinder parts*
> *Swim, and go smooth against the curling surge.*

(Statius, Achilleid I)

Among the many portrayals of hippocampus in statues, mosaics, tombs, coins, pottery, and jewelry, there is one particular Roman seahorse that caused an almighty stir when it was rediscovered in Britain toward the end of the Second World War. Engraved on a famous archaeological piece known as the Great Dish, a solid silver platter the size of a car hubcap, is a seahorse partaking in a busy scene of aquatic revelry and excess. In the center of the silver plate is the face of Neptune, or possibly Oceanus, a lesser-known Roman god of the sea, with dolphins leaping from his straggly seaweed beard. Around him a boisterous party is in full swing: Bacchus, the god of wine, presides over an exuberant celebration of music, dancing, and drinking; he is joined by a throng of naked sea nymphs who ride around on a collection of sea creatures, including a hippocampus. And among the scantily clad partygoers is the great hero Hercules, who has already had too much to drink and is having to be propped up.

The Great Dish, dated to the fourth century BC, was part of one of the greatest hauls of buried treasure ever found in Britain. The discovery of the precious silver was a tale of such intrigue and deceit that it still has people wondering what really went on. When it was reported in the newspapers in 1946, the story was so astonishing that it inspired Roald Dahl to write one of his few nonfiction works. Dahl, famous for his many children's fiction classics, including *Charlie and the Chocolate Factory* and *James and the Giant Peach*, claimed that when he read of the treasure's discovery in *The Times* (London), he leapt up without finishing his breakfast and drove straight to Suffolk to find the man who had found the treasure. His story, "The Mildenhall Treasure," describes in vivid detail the gray day in the harsh East Anglian winter of 1942 when Gordon Butcher was plowing a field at Thistley Green, a small village twenty miles northeast of Cambridge. Butcher was a farmworker and tractor owner who had been hired by another, richer man, Sydney Ford. At around three o'clock in the afternoon, as snow began to fall, Butcher's tractor struck something solid and he set to digging away the freezing soil to dislodge the plowshare. Expecting to unearth the tree root or stone that he had struck, he instead found the Great Dish. As Dahl describes it:

> He rubbed the rim with his fingers and he rubbed again. Then all at once, the rim gave off a greenish glint, and

Gordon Butcher bent his head closer and closer still, peering down into the little hole he has dug with his hands. For one last time, he rubbed the rim clean with his fingers, and in a flash of light, he saw clearly the unmistakable blue-green crust of ancient buried metal, and his heart stood still.

Butcher then made a big mistake. Not knowing what else to do, he went to find Ford and together they dug up not just the Great Dish, but an astonishing collection of thirty-four pieces of solid silver, including goblets, bowls, and spoons. Dahl implies that Ford was much more devious than Butcher and knew not only that wealthy Romans had lived in the area and may have left behind the treasure but also, since he hadn't actually found the haul himself, that he had no rightful claim over it. Under UK law, any item made from gold or silver that is dug up from the ground is called Treasure Trove and, no matter who owns the land where it was found or who dug it up, it automatically becomes property of The Crown. The vital piece of information that Butcher did not know was that because he had originally discovered the silver treasure, he could have handed the items over as Treasure Trove and claimed compensation from The Crown equal to the value of the haul. It would have made him a millionaire.

Acting fast, while the snow was falling, Ford assured Butcher that the dirty lumps of green metal were nothing more than junk and offered to take them away. Butcher, knowing no better, was pleased to be rid of them. And so, for the next few years, Ford kept the pieces, polishing them up to their full glory and gloating over the day, after the war ended, when he would make a handsome profit selling them to a private collector.

Dahl makes no attempt to hide his delight that Ford's stolen treasure was eventually discovered when a visiting archaeologist noticed a shiny spoon left carelessly on a mantelpiece. The author does, however, lament the fact that Ford got away with his deception and Butcher never received his million pounds—an unimaginable sum in those days, especially to a poor farmworker. Ford claimed in court that he believed the treasures were made of pewter, a metal not included in the Treasure Trove laws. The jury believed his story and, as a goodwill gesture, the "co-finders" were awarded one thousand pounds each by the British Museum in London, where the treasures remain on display today. One thousand pounds was also the sum Roald Dahl earned when he sold his story to the U.S. *Saturday Evening Post.* He sent half of his royalties to Gordon Butcher to thank him for sharing his story with the world. But even today, the tale of the giant silver plate, with its intricately carved seahorse, is far from over.

Ever since the Butcher-Ford court case concluded in 1946, academics have cast doubt on the stories they told, with various mutterings of conspiracy theories rattling through the world of archaeology. After the treasures were put on display, at great demand from an eager public, archaeologists from Cambridge University conducted a full excavation of the field at Thistley Green where Butcher and Ford had hesitantly claimed to have found the treasure. It came as quite a surprise when they found no other artifacts and no evidence of the hole that Butcher and Ford would have dug to retrieve the bulky silverware. And if Butcher's tractor really had collided with the Great Dish, surely it would have bashed a considerable dent in it? It is a little suspicious, then, that the dish on display in a glass cabinet at the British Museum is still in near-perfect condition. Not all researchers are skeptical, though. Archaeologist Richard Hobbs went to great lengths to figure out whether Dahl's story was plausible. He carefully pored over archives of the 1942 weather reports, hunting for any hints that would verify Dahl's account of snowstorms.

Hobbs even walked around carrying heavy buckets of sand to try and decide for himself whether Ford could have been strong enough to lift all the treasure in one go. Apart from a slight exaggeration of the inclement weather conditions, Hobbs found no fatal flaws in Dahl's version of events.

Other archaeologists have claimed that the treasure was in fact found in North Africa and that during a covert wartime operation, in which both Butcher and Gordon were involved, it was brought over by the air forces based at a nearby airfield. Hobbs pointed out that plenty of similar Roman silver has been dug up from the same region of South East England, so there is no need to dream up such intricate plots.[11] We may never find out the truth behind the mystery of the Mildenhall Treasure and the engraved silver seahorse, but Roald Dahl was right about one thing—it does make a good story.

As well as drawing their own images of half horse, half fishes, the Romans may have led to the appearance of seahorses in another branch of ancient European art a few centuries later. With their sights set on world domination, the Roman armies invaded England and continued to march north. When they reached the land now known as Scotland, they encountered an unruly scattering of tribes living in hill forts and farmsteads across the untamed moors and craggy glens. The Romans did not receive a warm welcome in the north and soon these so-called Scottish barbarians realized the importance of strength in numbers and united to form a formidable confederation to keep the invading forces at bay. The Romans nicknamed them the Picti, the "painted people," which may or may not have had something to do with a custom of decorating their skin in tattoos or war paint.

To this day, the Picts remain one of the most enigmatic of all the early European tribes. Other than a few items of metalwork and whittled bone, the only clues left behind of their existence are a puzzling set of intricate stone carvings that have been un-covered across northern Scotland; some of them are small frag-ments, some are enormous ten-foot edifices. They all take on a distinctive style and feature Pictish people wandering through a menagerie of splendid animals interspersed with inanimate ob-jects and geometrical designs, including interwoven knot work, an obvious Celtic influence. These works of stone art appeared out of nowhere, suddenly arriving in the seventh century AD with no obvious development or evolution in style. Then, after two hundred years and the carving of hundreds of stones, al-most as suddenly and for no apparent reason, the technique vanished.[12] Exactly why these carvings were created, who carved them, and what, if anything, they meant are all questions that still have no convincing answers. The messages left in the stone symbols are hotly debated by some archaeologists and aban-doned as meaningless conjecture by others. Were they totems to ancestral spirits? Did they denote family, rank, or status? Were they boundary markers or merely decorative works of art?

Whatever their intended purpose, the most alluring characters on the Pictish stone carvings are undoubtedly the animals— leaping salmon and sinuous snakes, prowling wolves and fear-some bears, elegant stags and waddling geese—many of them species that lived in northern Scotland and would have held some significance, symbolically or practically, to the Picts. A few of the carved creatures were clearly made up; the strangest of all was a four-legged character with a long curved snout that crops up on nearly sixty stones all across Scotland. Archaeologists named it the Pictish Beast or swimming elephant but are unde-cided about where the idea for it came from. The two-headed dragonesque brooches that were popular in northern England and Scotland in the first and second centuries AD could have been the inspiration for these odd-looking beasts. Another idea is that they were inspired by a marine mammal—a dolphin or a beaked whale—that would have visited the coasts and rivers of Pictish Scotland; the beasts' legs end in scrolling spirals, which could be a landlubber's idea of the submerged parts of a dolphin that rarely peek above the waves, and they are always carved leaning upward as if they are about to leap from the sea.[13] Pictish Beasts look something like fantasy writer Phillip Pullman's mulefa, a race of intelligent, four-legged creatures with dexterous trunks that appear in the stories of the His Dark Materials trilogy, trundling around a parallel universe using round seedpods as wheels.[14] And with their elongated tubular snouts and gently curved bellies, there is a temptation to link the strange Pictish Beasts to seahorses, but in fact there is no need to. Twelve of the Pictish stone carvings feature another character that bears an even closer resemblance to a seahorse.

The most convincing Pictish seahorse was discovered in 1986 on one of the Orkney Islands, which lie beyond the northernmost coast of Scotland. A farmer called Mr. Harcus was digging up potatoes in his field outside the small town of Tankerness on the East Mainland when his plow collided with something in the ground. In a discovery shrouded in none of the mystery and collusion of the Mildenhall Treasure, he hopped off his tractor, dug out the offending item, and brushed off its covering of soil to reveal a chunk of buff-colored sandstone into which was carved the distinctive outline of a seahorse, the length of two hand spans. Despite missing part of its head, probably lost when it was reused as a building stone, there is no mistaking the seahorse's down-turned snout, delicate arching neck, and curved body tapering to a tightly coiled tail tip. The body has no legs or wings or any other mythical adornment and is carved along its length with rows of deep ridges where in real life the seahorse's bony plates of body armor interlock. Across its face is the trace of a wry smile. This latest stone seahorse discovery now resides in the Tankerness House Museum in Orkney's main town, Kirkwall.[15] The rest of them are scattered across the northeast reaches of Scotland, some are carefully preserved in museums, others stand in churchyards and on roadsides where thousand-year-old artwork is being worn heartbreakingly away by the merciless Scottish weather.

Where, then, might the Picts have got the idea for carving these seahorselike creatures? Some archaeologists argue that they were surely inspired by the Romans, that Pictish stonemasons would have seen hippocampi prancing across Roman artifacts, many of which were brought north in deliberate attempts to foster relations with the northerly tribes. Indeed, several of the Pictish seahorses do have a distinctly classical appearance. The most elaborate example stands in the grounds of a small granite church surrounded by wheat fields in the tiny village Aberlemno, a short way outside the city of Dundee. Among rows of Christian gravestones looms an imposing Pictish carving, the size of a single bed standing on its end. On one side is a bloody battle scene, an unusual subject in Pictish art, with soldiers on foot and horseback fighting with swords and spears; in one corner a raven pecks at a soldier's dying body. The stone is thought to commemorate the Pictish victory over the Angles at the Battle of Nechtansmere in 685 AD at a site ten kilometers away. On the other side of the stone is a vast Celtic cross filled in with intricate knot work and surrounded by various intertwining beasts, including on the bottom right-hand corner a pair of exquisite symmetrical seahorses. Their heads are clearly those of horses, with long muzzles, little pricked-up ears, and flowing manes. Lying nose to nose with forelegs raised, the pair touches hooves, perhaps in battle. Their long bodies weave sinuously together; in place of rear legs, each one has a tall dorsal fin and a sickle-shaped tail fin like a tuna. A Celtic triangle fits neatly in the space between their tails.[16]

Many archaeologists have described some of the other Pictish seahorses as "dog-headed fish monsters,"[17] including John Romilly Allen and Joseph Anderson in their seminal 1903 tome, *The Early Christian Monuments of Scotland*. Allen and Anderson are convinced these creatures are imaginary concoctions, perhaps a maritime symbol intended to represent the barking of the waves. It certainly is difficult to ignore the canine qualities of some of these carvings. On the grounds of Brodie Castle, a few miles outside Inverness, two seahorses stand on guard, their mouths wide open displaying rows of fearsome teeth, far more terrifying than any real, toothless seahorse.

It has been suggested that at first the Picts carved seahorses in the Roman style, as seen at Aberlemno, and that over time they morphed these creatures, eventually creating the dog-headed fish monsters.[18] But there is one hitch in that theory. Since the end of the nineteenth century, a tiny schoolhouse in the Scottish village of Meigle has been home to a collection of twenty-seven Pictish stone carvings that were found in the nearby churchyard and built into walls of the old church. On one of the stones are two seahorses that gaze in contemplation at each other. The first has forelegs and a fish's tail and resembles other dancing stone seahorses kept at the National Museum of Scotland in Edinburgh. The second is smooth, curved, and legless, more like the seahorse at Orkney. Either the theory of evolving Pictish seahorses is incorrect or this mismatching pair was a deliberate combination of new and old designs.

Is it possible that the Picts could have seen real seahorses for themselves and decided to carve them in rock? One archaeologist, Antony Jackson, confidently postulates—but not without criticism—that this was certainly the case. He is convinced that the Pictish animal carvings were indeed lineage totems and that the seahorses, along with other water-based creatures, represent dominion over the seas and rivers, perhaps signifying families who were great fishermen or sailors.[19] Isabelle Henderson, a life-long expert on the art of the Picts, posed the idea that pairs of seahorses could have been symbols of guardianship with a protective role and that they were peaceful benign creatures, but with no specific links to real species.[20] Either way, it is feasible that the Picts would have seen real seahorses. Maybe the valuable dried seahorses used by Romans as medicines could have been traded as far north as Scotland. And it may seem unlikely, but today seahorses are known to live along the Scottish coasts, raising the possibility that the Picts may have snagged wild sea-horses in their fishing nets or seen them washed up in a tangle of high-tide flotsam. Thanks to the warm current of the Gulf Stream, lapping across the Atlantic from the Caribbean, the northwest coast of Scotland is just warm enough to be home to long-snouted seahorses.[21] On a map of their global distribution, there is even a small dot in the Orkney Isles and another one far-ther north in the Shetland Isles, marking the northernmost edge of their range, although sightings are extremely rare; the last time fishermen in Orkney found seahorses hooked in their nets was at the end of the Second World War. Since then, only a few dead ones have washed up on beaches, but perhaps that was all the Pictish artists needed to gain inspiration for their carvings.[22]

Whether or not real seahorses were the impetus for the Pict-ish seahorse carvings, the idea of watery horses did not die out when this northern tribe disappeared, absorbed perhaps into the Gaelic Kingdom of Alba or overtaken by the Vikings from the north. The seahorse, with the head of a horse and tail of a fish, lived on into medieval times and beyond. It was around this time that volumes of animal stories known as "bestiaries" or "books of beasts" became extremely popular, not only among Christians in Western Europe but also throughout the Islamic world in North Africa and the Middle East. The lavishly illustrated books featured catalogues of animals, some of them real, such as hyenas, beavers, sheep, lions, and hedgehogs, and others imaginary, including unicorns, basilisks, gryphons, and manticores. Each creature came with a story that offered a moral and usually religious lesson stemming from beliefs that all living things, being God's creations, embodied their own spiritual meaning.[23]

The earliest known bestiary, called the *Physiologus*, came from an anonymous second-century Greek writer who amalgamated many animal stories by earlier classical scholars, including Herodotus, Aelian, and Pliny the Elder. Scores of later versions reworked and expanded on a Latin translation of the original text. The well-preserved manuscript of the thirteenth-century *Aberdeen Bestiary* tells us about an animal called the ibex, which "has two horns of such strength that, if it were to fall from a high mountain to the lowest depths, its whole body would be supported by the two horns."[24] This creature, the book continues, "represents those learned men who are accustomed to manage whatever problems they encounter" and who "supported as by two horns, they sustain the good they do with the testimony of the readings from the Old and New Testament." The mightiest of all the beasts was undoubtedly the lion, whose character traits were repeatedly held up for divine comparison. It was long believed lionesses gave birth to dead cubs that she watched over "until their father comes on the third day and breathes into their faces and restores them to life."[25] Thus, in the *Aberdeen Bestiary* we are told, "the Almighty Father awakened our Lord Jesus Christ from the dead on the third day."

Among these religious writings, there was an enduring conviction that for everything that lived on land, an equivalent could be found in the sea: a reflection of the Creator's plan for "perfect symmetry." Pliny the Elder had laid down this idea a long time previously when he wrote it "may very possibly be true, that whatever is produced in any other department of nature, is to be found in the sea."[26] He asserted that in the oceans there were "forms not only of terrestrial animals, but of inanimate objects even" with such creatures as "the grape-fish, the sword-fish, the sawfish and the cucumber fish, which last so strongly resembles the real cucumber, both in color and smell." The pages of many ancient bestiaries depict all manner of land creatures embellished with fish tails: not only sea horses but also sea bulls, sea goats, sea lions, and so on. A fifteenth-century encyclopedia of the animal, plant, and mineral kingdoms, called *Hortus Sanitatis*, included several intriguing woodcuts of seahorses, one caught in a fisherman's net, as well as a dog fish and a pig nestled inside a seashell.

Belief in gods of the sea was something else that continued long after the breakup of the Roman Empire. In Celtic folklore, Manannán mac Lir was god of the seas, and under various guises appears in Irish, Scottish, and Welsh legends. Just like Poseidon, he rode across the oceans in a chariot pulled by great horses, the most famous being Splendid Mane, who was said to be swifter than spring wind. The Isle of Man was named after this sea god, and fishermen would say they could see Splendid Mane galloping along cresting waves across the Irish Sea, where Manannán mac Lir presided deep down on his conch shell throne.[27]

Many legends tell of water spirits that take the form of horses. Kelpies are supernatural beings that supposedly haunt rivers and freshwater lochs in Scotland and Ireland. If you mount a kelpie, it will leap into the water and try to drown you. Similar malicious beasts were called nuggles in the Orkney Isles, shoopiltees in the Shetlands, and across Scandinavia stories were told of the bäckahästen, or the brook horse, also known as nykur. The Each Uisge is an even nastier water horse that inhabits sea lochs in Scotland. It hangs around the water's edge, tempting passersby to hop on for a ride. Those who do will be stuck fast by a layer of gooey glue before being plunged underwater and voraciously devoured. All that remains of the Each Uisge's victim is the liver, which floats up to the surface.[28]

And today, thousands of years after the first mythical seahorses were dreamed up, the idea of water horse spirits may even have spawned a legend that many people still want to believe: the Loch Ness Monster.

By the time the skies began to darken over the Uşak desert on the sixth of June 1966, Osman and his accomplices were still carrying away armfuls of ancient treasure from the newly opened catacomb. Following a lead from a rival villager, who was perhaps envious that he had not scored any loot for himself, the police finally arrived at the gravesite and were greeted by an outburst of gunshots as the tomb raiders ran off into the night. But the curse of the tomb apparently caught up with many of the looters. Osman endured a lengthy jail sentence, while several of the others met with early deaths, and one was blinded. The police managed to recover some of the treasure, which was put on display at the Museum of Anatolian Civilizations in Ankara, but by 1970 the remaining items had found their way to the United States. Over three hundred pieces of gold, silver, and bronze, which came to be known as the Lydian Hoard, were smuggled across the Atlantic by North American traders who sold the treasure to the Metropolitan Museum of Art in New York for one and a half million dollars. The museum kept suspiciously quiet about the Mediterranean jewels, including the golden seahorse, locking them away in the basement where, once again, the Lydian Hoard lay quietly in the dark. The artifacts weren't listed in the central catalogue, and looking back, the museum acted very much like it was hiding a guilty secret.[29]

Little was seen or heard of the Lydian Hoard for over a decade. It might have lain in obscurity for many more years were it not for two Turkish journalists, Özgen Acar, a veteran journalist with a keen interest in archaeology, and Melik Kaylan, a young Turkish writer working in New York, who together set out in 1987 to investigate what had happened to the Lydian treasures. They had received a tip from Thomas Hoving, former director of the Metropolitan Museum who had been appointed shortly after the Turkish treasure arrived. In his autobiography, *Making the Mummies Dance*, Hoving revealed his suspicions about the Turkish artifacts. He describes a meeting where he warned his staff that the museum would eventually have to return the dubiously acquired treasure. "We took our chances when we bought the material," he told them.[30]

Acar and Kaylan set up a covert mission into the Metropolitan's vaults to photograph the missing treasure.[31] Back in Turkey, the journalist duo tracked down some of the villagers who had broken into the tomb, including Osman, and showed them the photographs from the United States. Despite the passage of time, the surviving tomb raiders immediately recognized some of the pieces they had taken from the Uşak desert. And so, with these new eyewitness accounts, the Turkish government finally had all the evidence it needed to demand its treasure back from the other side of the Atlantic.

In New York, the Metropolitan Museum refused to play ball. Claims were made that, according to state law, the Turkish government had already missed the three-year deadline to take action. It took another three years for the museum's plea to be overturned, when the judge overseeing the case agreed that the true identity and whereabouts of the treasure had only recently come to light; the three-year clock had started ticking only when Turkey first demanded the return of their artifacts. What followed was a million-dollar lawsuit, which eventually ended in 1993. The museum's last-minute attempts to negotiate a fifty-fifty split failed, and they were forced to admit they had known all along that the fabled items had been stolen. All three hundred and thirty-six pieces of the ancient treasure, including the golden winged seahorse, were sent back to Turkey in what should have been a happy ending to the story of the Lydian Hoard. But sadly, it didn't end there.

For eight years the treasures were put on public display at the Uşak Museum in Turkey, not far from where they had originally been found decades earlier. Then, in May 2006, the Turkish news was rocked with the headline that the golden winged seahorse brooch on display was not a piece of ancient treasure from two and half thousand years ago, but was in fact a twenty-first-century fake. The scandal raced around the world and soon accusations were flying that dozens of other artifacts from the Lydian Hoard had been replaced with counterfeits during a major inside operation coordinated by the director of the Turkish state museum, Kazim Akbiyikoglu. Along with eight other museum officials, Akbiyikoglu was arrested on suspicion of collusion and theft. The case of the Lydian Hoard had been quietly reopened five months earlier when the governor of Uşak received a letter claiming that the golden seahorse had been stolen. Close inspection of the brooch on display at the museum exposed it as an inaccurate replica, weighing slightly more than it should have.

Evidence in the Lydian Hoard trial was incontrovertible. Tapped phone conversations were heard in court, with Akbiyikoglu and collaborators referring to the original brooch as an "organic tomato" and the fake as an "inorganic tomato." Photographs were found on a mobile phone showing the original brooch laid out on the knee of one of the accused, who wore the same pair of distinctive trousers when he was brought into custody; the offending trousers were used as evidence in the trial. And despite claims that he realized some of the artifacts were fakes when they were returned but had kept quiet to save the Americans' reputation, Akbiyikoglu was sentenced to twenty-five years in jail, along with eight members of his staff who each got eight years.[32]

It was all a colossal embarrassment for the Turkish government, who had fought so hard for the return of the Lydian antiques. And the story has stoked the fires of international debate over the rightful ownership of ancient objects. The history of archaeology, treasure hunting, and exploration is plagued with accounts of ancient relics being procured from their original locations, often in the less-developed corners of the world and often in dishonest circumstances, and finding their way into the museums, galleries, and private collections of the West. The case of the Lydian Hoard supported the distasteful notion that lower-income countries are perhaps more prone to corruption and might not be the best places for safeguarding priceless treasures. In 2006, the Turkish culture minister, Atilla Koc, stated in *The Times* newspaper in London, "You can secure the outside of the museum as much as you want, but no tool has yet been invented to secure the people on the inside."[33] The case has undoubtedly set back the efforts of many archaeologists who campaign for the repatriation of other ancient artifacts to their countries of origin, including the British Museum's Elgin Marbles, a collection of Greek sculptures that many believe should be returned to Athens.

The only clues to the current whereabouts of the golden winged seahorse are rumors that the buyer who spearheaded the covert operation was Japanese. For now at least, the golden seahorse once more lies somewhere hidden, although maybe, one day, like so many other buried works of ancient seahorse art, it will reappear.

Did seahorses really inspire all these works of art and all these stories? Did they really lead people to sculpt so many beautiful and valuable objects that stirred up scandals and sparked such jealousy and greediness? Just one tiny creature? Well, of course, it was not just one tiny creature, but millions of them, living their hidden lives in shallow seas all around the world. And to see one, to catch a glimpse and snatch an idea, all these artists, jewelers, mythmakers, and storytellers did not actually need to see a seahorse for themselves, alive and swimming through the seas. It just so happens that so much of the seahorse's magical appearance lives on after death. A seahorse's tough outer skeleton leaves behind an elegant corpse, a lasting impression of their winsome proportions, their intricate spines and ridges, fairy-tale crown, and curling tail. On my desk, in a small, gray cardboard box, rests a tiny dried seahorse as tall as my little finger. It has been dead for at least thirty years, a gift from a friend whose mother bought it for her when she was a little girl, and yet after all that time, it kept its delicate features. Similar long-gone beings lie scattered along shorelines, the vivid memory of so many baby dragons, waiting to be picked up by a passing beachcomber, slipped into a pocket and passed on to a friend. And for almost as long as people have been capturing seahorses in paintings and carvings, stories and myths, there have also been people who stood for a moment longer on the beach, holding on to a tiny dried form, pondering what kind of creature it really is and where it came from.

Chapter Two

AN A-Z OF SEAHORSES

A native tribe of Mexican Indians called the Seri tell a story, handed down from their ancestors, of how the seahorse came to be. In the beginning when the world was new, all the animals could talk and wore clothes like people. The seahorse lived on Tiburón Island in the Gulf of California and was a fat, well-fed fellow, not the skinny creature we know today. He was also a prankster and a schemer and one day did something unmentionable, something that Seri folklore has since forgotten. Whatever his misdemeanor, it incurred the wrath of all the other animals and a great fight ensued. Finding himself outnumbered, the seahorse tried to escape while rocks and stones were being hurled at him,

꩜

Above: Young seahorse viewed from side as a transparent object

Credit: Theodore Gill, 1905

cutting and bruising him as he fled. Eventually, he reached the beach and, with nowhere else to run, he tucked his sandals into his waistband and dived into the sea, never to return. To this day, whenever you see a seahorse, it is scrawny and fleshless, with a tiny fin in the same place where its outcast forefather stowed his sandals before abandoning the land forever.[1]

Like the mythology of many ancient cultures, this Mexican legend embodies part of a storehouse of knowledge accumulated through generations to help explain and understand the mysteries of life. The Seri told stories of animals they encountered in their daily lives as hunters and fishermen; occasionally they would have caught seahorses in their nets or found them washed up on a beach. Their stories were often deeply allegorical and laid down which behaviors were acceptable and which were not. While this story may not help explain how seahorses truly came to be, whatever social blunder he committed, a young Seri would have thought twice about doing the same.

For many centuries, people marveled at the seahorse's eccentric profile, wondering why they look the way they do, where they came from, and exactly what sort of animal they really are. Long before their true identity was revealed, seahorses were assigned the name Hippocampus after the character in ancient Greek and Roman myths. Some think the *campus* ending comes from an ancient Greek word meaning "caterpillar," which makes sense when you look at the seahorse's narrow tail covered in rounded spines that could easily be mistaken for rows of caterpillar feet. Seahorses look so magical and unlikely, who can blame medieval merchants for declaring that they had discovered baby dragons from faraway lands? As recently as the nineteenth century, Victorian scientists and naturalists were still arguing over whether to classify seahorses as a type of insect, an amphibian, or perhaps an odd-looking crustacean—a prawn or a shrimp with no legs.

It is now, however, agreed that seahorses are a type of fish, an indisputable diagnosis based on their having a pair of gills to breathe through and an internal airbag called a swim bladder to control their buoyancy underwater (human scuba divers wear a Buoyancy Control Device, or BCD, to achieve a similar effect). Narrowing things down, seahorses are bony fish (or teleosts), since beneath their hard outer casing of interlocking plates lies a tiny articulated spine stiffened with bone, distinguishing them from the sharks and rays that instead support themselves with softer cartilage.

Seahorses are marine fish, unable to tolerate freshwater, although some do sneak into estuaries and rivers in places still touched by the salty lick of ocean tides. Their favorite stomping grounds are a trio of marine habitats: the cosmopolitan coral reefs, verdant lawns of seagrass, and the twilit roots of mangrove forests that fringe tropical shorelines. Other popular seahorse hangouts are gardens of sponges and seaweed; some like to crouch in sandy hollows on the seafloor, and some live happily alongside people, sheltering between the pilings of jetties or looping their tails through nets used to keep sharks away from swimming beaches.[2] There is even one unusual seahorse from Hawaii that drifts through open ocean. When they are fully grown and stretched out straight, seahorses range in size from around two centimeters, the width of your thumb, to more than thirty centimeters. The biggest, the big-belly seahorse from Australia and New Zealand, grows to be as long as your forearm from wrist to the crook of your elbow. Seahorses can live for several years, the larger ones living the longest, up to five or even ten years. They are not the most social of creatures; they don't congregate in seahorse herds, instead spending much of their lives alone. The most popular pastime for a solitary seahorse is eating. Their preferred meal is a handful of microscopic shrimp, part of the zooplankton that waft through the oceans like grass seeds blowing in the wind.[3] And when the sun sets and waters darken, the seahorse, like many fish with no eyelids, enters a dormant, wide-eyed, sleeplike state.[4]

Seahorses share many of their odd features and unusual habits with the rest of their family, a group of fish called the Syngnathidae (from Greek words *syn* meaning "with" or "together" and *gnathos* meaning "jaw"). The oceans are home to more than two hundred species of syngnathid, an idiosyncratic bunch. Nearly all of them are pipefish, which look like play dough seahorses that have been rolled out straight and thin, sometimes with a little fan of a tail stuck on the end. Pipehorses and pygmy pipehorses are vanishingly rare creatures that are stuck somewhere between a seahorse and a pipefish. They look like seahorses that have been on a diet and learned how to swim stretched out and horizontal, using just the very tip of their tail to hold on to things. The two species of seadragons are the most outlandish syngnathids of all. These flouncy fish are like seahorses that were invited to a fancy dress party and made an extra special effort with their costumes. Leafy seadragons are naturally festooned in elaborate outfits of green ribbons and streamers, outgrowths of skin that create the near-perfect illusion of a tangle of seaweed, making them as invisibly camouflaged as a woody stick insect. Weedy seadragons are slightly less outrageous with fewer dangling ornaments, but still they put on an eye-catching display, with banana yellow freckles and electric blue racing stripes. The Syngnathidae family has several strange cousins, including the seamoths that do insectile impersonations on the seafloor and the trumpetfish, flutemouths, and tubesnouts, which all look a little like giant pipefish with elongated, pointy snouts. Then there are the quirky ghost pipefish; some are uncanny duplicates of detached, drifting seagrass leaves, and others resemble fiery red, stringy sea sponges or spiky feather stars; there is even an orange, hairy ghost pipefish that looks like a monster from the *Muppet Show* and has to be seen to be believed.[5]

Despite their bewildering appearance, there is no doubting that seahorses, and their syngnathid relatives, are indeed bony fish and not insects or shrimps (or dragons). Nonetheless, they don't look much like any of the other fish that roam the seas. So why do they look the way they do?

One of the first times the image of a seahorse appeared in print was in 1565, in an edition of the *Commentaries on the Materia Medica of Dioscorides* by Matthioli, an ancient list of traditional European medicines. Matthioli included a detailed and realistic woodcut print of two common Mediterranean species, one with a bulging belly, the other with an unkempt mane of straggly hair.[6] These early seahorses display the suite of unique characteristics that set them apart from all the other fishes. The curious elegance that we so admire is not simply the fortuitous result of an arbitrary collision of features. Instead, each ingredient has a vital role to play in sculpting a form perfectly adapted to a certain way of life. Seahorses look the way they do because it works. And as we could, if we wanted to, learn how a toaster works by taking it apart and figuring out what each component does to help produce a crispy browned slice of bread, so we can do the same for seahorses—although, of course, not literally. Spend a short while watching a seahorse and it soon becomes clear how each apparently baffling physical component lends itself to the sedate seahorse way of life spent hidden from inquisitive eyes.

One of the most distinctive attributes of a seahorse is its tapering appendage, taking the place of the vertical fanlike tail that most fish flip from side to side to push themselves through the water. Instead of using their tails for swimming, seahorses sit upright and use their tails for hanging around. When a strong current tries to sweep a seahorse away, it will wrap its tail around a seagrass blade or coral finger, clinging on tight, swaying to and fro with the rhythm of the surge. A common pose for a seahorse—when the underwater wind is not blowing—is with its tail rolled forward in a tight coil, resting snugly beneath its belly. Perhaps it is more comfortable that way, the seahorse fetal position perhaps? And yet a seahorse tail is a most supple device: It can bend in any direction or stretch out straight as a ruler. When they feel like it, seahorses can scratch their own heads and use their tails as a neck scarf.

Although they have no tail fins, seahorses are not entirely fin free. They have four of them—tiny transparent things. The largest lies at the base of the tail—where the Seri Indian seahorse tied his sandals—and there is one behind each cheek, which from the front look like sticking-out ears, plus a minute protrusion under the belly. These fins are held out stiffly with an array of thin bones called rays and are waved in undulations like the falling slats of a picket fence, up to forty times a second.[7] Seahorses beat their fins so quickly they become an invisible blur to the human eye. Such diminutive fins may not permit great speed, but that doesn't matter for these placid creatures, accustomed as they are to living at an unhurried pace. As Jacques Cousteau once wrote, "These slowpokes may take one-and-a-half minutes to cross a one-foot area."[8] Where speed is forsaken, maneuverability is gained. Using their four fins, seahorses can move and hover in any direction they choose, just like a helicopter or a dragonfly: up, down, forward, backward. But all that swimming is hard work, and the seahorses have an almost uncontrollable urge to hold on to the nearest thing they can find.[9]

To help them survive life in the slow lane, seahorses come equipped with a special skill: camouflage. When a hungry predator approaches, seahorses can't swim off and disappear in a flash but instead they disappear right where they are in a skillful game of hide-and-seek. Seahorses are wily masters of disguise; a shiver of an eye may be all that gives them away. In place of the customary coating of fish scales, seahorses have an armored suit of interlocking bony plates covered in a layer of color-changing skin. They share the octopus's clever trick of coordinating seamlessly with their surroundings. They can mimic the fallen yellowing leaf of a mangrove tree or a bright orange coral. A little like seadragons, although not as showy, seahorses can sprout weedy filaments and protrusions to help them hide in a patch of seagrass or a knobbly sponge or among the pink nubbins of a sea fan. The sturdy suit of armor is itself good defense against would-be predators, making seahorses hard to swallow. As the great nature writer Rachel Carson wrote, it is "a sort of evolutionary harking back to the time when fish depended on heavy armor to protect them from their enemies."[10] It is because of their unyielding "exoskeletons" that seahorses were for a long time mistaken for crustaceans or insects. However, unlike a lobster or a beetle, the seahorses' armor expands as they grow, giving them no need to periodically molt and cast off their undersize garments.

Something else that sets seahorses apart from all their fishy cousins is their horselike head, with puffed-out cheeks and elongated snout topped off with a pair of swiveling turrets for eyes. With their unusual vertical stance and head bent over at ninety degrees, seahorses are the only fish that have necks. It gives them just the right shaped head for seeking out and guzzling down their shrimp prey, the same microscopic source of nourishment that fuels the world's largest fish, whale sharks and basking sharks, as well as the most colossal creatures to ever roam the planet, the blue whales. These giants cruise the open oceans, mouths gaping wide, straining gallons of water for thousands of minute crustaceans every minute. But living a sedate life on the seafloor, and with a more reserved appetite, seahorses adopt a wholly different strategy: They sit and wait for food to come to them. Like a chameleon's, a seahorse's eyes can move independently of each other, essential for spotting specks of potential food floating all around them in the water column. West African folklore warns us to fear chameleons because it is said they can look into your future and into your past at the same time.[11] Perhaps the same goes for seahorses?

When a seahorse spies a shrimp drifting into range, it smoothly lines itself up, preparing for action, nose turned down, shrimp just above. Then, with sudden immense speed, the snout snaps upward and the seahorse inhales a vortex of water, drawing the hapless shrimp into the narrow toothless mouth. Some say that a seahorse's cheeks generate such a powerful suck that smoky puffs of pulverized food escape through their gills, adding fuel to the myth that they are indeed fire-breathing dragons.

There may be a lot of sitting and waiting in between, but when they get going, seahorses are very fast eaters. High-speed cameras have watched them gobble their food in under six milliseconds, putting the seahorse gulp among the fastest-known movements of all the vertebrates. Theirs is a specialized technique known as pipette or pivot feeding; imagine trying to eat a bowl of Rice Krispies at breakneck speed using nothing but a drinking straw. The seahorses, along with all the other Syngnathidae, are well equipped to make such rapid and accurate movements. Studies have shown that for fishes feeding on small, fast-moving prey, the most efficient way of hunting is to have a long tube for a snout; all it takes is a slight turn of the head to bring mouth into contact with distant morsel. Among all the pipefish, the species with the longest snouts are the ones that pursue the speediest prey.[12]

The seahorses' unconventionally shaped heads may also hold the secret of their ability to communicate with one another. In his 1905 paper "The Life History of Sea-horses," Theodore Gill included a description of a "noteworthy peculiarity" mentioned twelve years earlier by another scientist named Kent. He was sketching seahorses inside two glass jars when "unexpectedly a sharp little noise was heard at short and regular intervals to proceed from one of the vases placed on a side table, and to which a response in a like manner was almost immediately made from the vase close at hand." Exactly how seahorses make chirping noises like grasshoppers and snapping sounds like a finger click was a long-running question that vexed many scientific minds. In 1874, Dufossé described quivering in a seahorse's lower jaw that was too fast for the human eye to see but was "appreciable to touch" and "rarely distinctly audible."[13] More recently, high-speed video and audio recordings have revealed how a bony ridge at the back of the skull rubs against the crownlike coronet, creating the sounds of seahorses.

And so, after our tour of the seahorse's odd physique, we are left with just one feature to explain: the protruding pouch, borrowed perhaps from a kangaroo or a wallaby. And this is where we discover far stranger goings-on in the seahorse realm than simply their unconventional anatomy.

Right now, somewhere in the world, early-morning sunbeams pierce through shallow water like spokes of a wheel and cast quivering pools of brightness on the seagrass meadow below. The night shift has ended and diurnal creatures begin to emerge from sleeping hideaways: grazing rabbitfish, bucktoothed parrotfish, and feisty damselfish tending their farms of algae. Two tiny silhouettes come together like a pair of knights on a chessboard. The seahorses greet each other with a nose-to-nose caress and, wrapping their tails around a single blade of grass, they begin a seductive dance, spiraling round and round each other. Blushes of orange and pink give away their emotions and for a moment the seahorses let go of their holdfast and swim together, heads tucked down, tails entwined. A gentle humming and clicking from the male is the sound track to their flirting.[14]

The first time a seahorse couple meet, this gentle courtship carries on for hours, days even, and it is a risky time. Driven by hormones that interfere with the instinct to hide, they abandon the camouflaged safety of their seagrass home. The female initiates sex by reaching up toward the surface, pointing skyward with her snout, stretching her body as straight as it will go. This proves quite irresistible to the male, who immediately responds by pumping his tail vigorously up and down. It is an awkward motion, so unlike his usual slow, graceful movements, but he continues gyrating as he trails after her, obsessed. The couple halts in the open water column and hold their bodies close, forming a heart shape with their touching snouts and bellies. Their first attempt isn't quite right, so they break apart and try again several times until their position is perfected, the female just above the male. Then an extraordinary thing happens. A short hollow tube emerges from the female, which she pushes into an opening in her partner's belly. The couple raise their heads and arch their backs as the female shoots an egg-laden liquid into the male. Copulation is perfunctory, taking just six or seven seconds. When the male is full with a precious cargo, he wanders off, his bright mating costume already fading. He sways and wiggles his body, settling the eggs into position where they will remain for the next few weeks, growing in a protected internal pond.[15]

The strangest thing about seahorses is that male seahorses are the only males in the world who have experienced—firsthand—the agonies of childbirth. Apart from unlikely Hollywood movies, males never get pregnant. Except, that is, for seahorses.

Admittedly, there are many fathers who do a great job of helping out with the youngsters. In eastern Australia's rain forests, tadpoles of the marsupial frog wriggle into special pouches slung on their dad's hind legs. Six weeks later, out hop the next generation of fully formed miniature frogs.[16] These, and many other caring males—including the pipefishes and seadragons—undoubtedly deserve praise for their efforts, but only male seahorses become truly pregnant, nurturing their young inside their bodies, providing them with food and oxygen, whisking away waste products. This is all the more remarkable when we consider that pregnancy is a rare occurrence in fish, even among females.

Nowadays, when people first hear about seahorse males getting pregnant, the question that naturally follows is "so what makes them male, then?" Quite simply, the answer is sperm. The fundamental and unswerving difference between males and females of any species is the way they pass on their genes to the next generation. Ladies produce large sex cells (or gametes) that are immobile and expensive on energy; gentlemen, on the other hand, make gametes that are small, mobile, and above all cheap. A choice between round eggs or tadpole-like sperm is essentially all that separates woman from man, doe from buck, mare from stallion, and so on.

Despite such a clear definition of "male" versus "female," it took marine biologists a long time to understand what was

going on with seahorse sex. It was the ancient Greek philosopher Aristotle in the third century BC who first wrote about the
unusual reproductive habits of the Syngnathidae family. In his
book *On the History of Animals*, Aristotle went into extraordinary detail on the lives of many fish species that he encountered
while staying on the Mediterranean island of Lesbos. Among
his descriptions is an unmistakable account of the spawning behavior of the needlefish known also, he wrote, as the pipefish:

> The pipefish, as some call it, when the time of parturition
> arrives, bursts in two, and the eggs escape out. For the fish
> has a diaphysis or cloven growth under the belly and ab-
> domen (like blind snakes) and after it has spawned by the
> splitting of the diaphysis, the sides of the split grow to-
> gether again.

There is little doubt that Aristotle must have seen real pipefish, but he clearly did not suspect that the male pipefish rather
than the female was the egg carrier, for once they had hatched:

> the young fish cluster round the parent like so many young
> spiders, for the fish spawns on to herself.

It wasn't until the eighteenth century that scientists realized
something strange was going on and began to study syngnathid
sex in detail. For four decades, arguments flared over which sex
carries the eggs during a seahorse or pipefish pregnancy. Everyone agreed that the females produce the eggs, but it wasn't clear
whether or not they handed them over to their male partners to
look after. The academic tussle was played out on the pages of
specialist journals, until the debate was finally laid to rest in the

1870s when several scientists observed pairs of seahorses engaging in tight embraces within the confines of the laboratory. Those watching closely enough witnessed the transfer of eggs from female to male.

When it comes to making eggs or sperm, there is no avoiding the expensive versus cheap argument: Females of most species make a limited number of eggs and tend to look after them well as they develop, while males make torrents of sperm. This means that pregnancy isn't usually a great option for males. Why should a male adopt the most female of traits and spend time looking after a single brood of young when he could be roaming around, scattering his abundant sperm, fertilizing many more broods elsewhere?

To start with, a large proportion of male fish have an inclination to care for and nurture their young. The oceans, lakes, and rivers of the world are full of males that spend time and energy looking after their offspring; they jealously defend nesting territories, guard fertilized eggs, and even carry them around in their mouths or glued to their bodies. But what makes male fish such great dads? The beginning of an answer is the lack of obvious choice between male and female fish for who is best suited to the job of child care. For mammals, there is only one contender; the female nurtures young inside her womb, leaving the male pacing about with few options to help except fending off predators from mother and babies and bringing them food. Female fish, however, usually carry out external rather than internal fertilization and can benefit from abandoning their eggs and concentrating on feeding, so that next time they can make bigger, better eggs. On the other hand, male fish can boost their credentials by hanging around and watching a nest of eggs. By claiming and defending a piece of prime territory, a male can look after several clutches at once and in doing so he becomes irresistible to the ladies who prefer responsible caring types to father their children.

For further clues, we can cast a glance back in time. Before seahorses became seahorses, they looked something like their modern-day relative, the stickleback, a typical silvery fish that lives in streams and rivers in Europe and North America. Today, stickleback males nurture nests of eggs and it is likely that the seahorses' ancestors did the same thing. But for some reason, somewhere along the line, the males evolved to carry their eggs around with them like a treasured collection of marbles; maybe nest sites were in short supply or perhaps a surfeit of predators meant offspring needed extra protection and couldn't be left lying around on the seabed. Either way, from this egg-carrying ancestor it is easy to see how the seahorse's pouch and male pregnancy evolved, because the pipefishes, those pencil-straight seahorse cousins, elegantly demonstrate all the in-between steps from smooth belly to secure pocket. If we sort through the many hundreds of pipefish species, we can divide them into groups that have stopped off at various stages along the evolutionary pathway to male pregnancy. First we see species like the flagtail pipefish, which in many ways resemble the seahorse's early ancestors; they swim freely in the water column, have no pouch, and simply stick eggs to their bellies. Then we see yellow-scribbled pipefish with a flap of skin that they use to grasp partially on to the egg clutch. Then there are many species, including chocolate pipefish that, like seahorses, have slipped down from open water to live on the seafloor amongst crawling nibbling predators. By this stage, the skin flaps have fused to form a protective hollow tube, which helps to shield offspring from many hungry mouths. Only in the seahorse has that tube sealed off at top and bottom to form a fully enclosed pouch with just a small opening to let things in and out.

It has long been assumed that male seahorses squirt sperm directly into their pouches after the female has transferred her eggs—since that would make the most sense—but recent close inspection of seahorse anatomy has revealed something else is going on. It turns out that such a simple method of fertilization is impossible because the sperm duct, the opening that sperm emerge from, is located outside the male's pouch. They have "aquasperm" that shoots into the sea instead of "introsperm" that stays safely inside either male or female.[17] Such an apparently inefficient design increases the chance that sperm will get swept away before the male has a chance to suck them up again. But it just goes to show that the seahorse's pouch wasn't deliberately designed from scratch but in a series of gradual, blind evolutionary steps.

And so the marsupial pouches of male seahorses are no longer a mystery to us but instead they are logical places for young seahorses to grow up. This does, however, leave us with one more surprise. When evolutionary biologists discovered that male seahorses become truly pregnant, they rubbed their hands in anticipation; it gave them a perfect opportunity to test out their theories of how differences between the sexes evolve. They expected to find the females, now unshackled from the toils of pregnancy, had kicked up their heels and adopted a typically male habit, spreading their gametes as far and wide as possible. But no, most female seahorses are loyal to one male throughout his pregnancy and do not mate again until he is ready. Many seahorses are monogamous throughout whole breeding seasons, returning to the same partner time and again. Some may even stay in devoted couplings for much of their lives. So what benefits do females gain by abandoning pregnancy while at the same time sticking with one mate? The answer could lie in their natural rarity. Seahorses don't live in crowded neighborhoods, possibly because their plankton food is too scarce to support more than a handful of adults in a patch of habitat the size of a tennis court. And with such limited social opportunities and meager swimming skills, seahorses can't rely on roaming around looking for a new partner every time they are ready to breed. As soon as they have found a suitable mate, it pays off in the long run for both males and females to stay together. If males are unlikely to find a profusion of other mates, it isn't a huge sacrifice after all to settle down, be faithful, and become pregnant. And taking on the reins of pregnancy gives male seahorses one last added benefit: full reassurance that all the babies he's caring for are definitely his own, something other males, most notoriously human beings, can't be absolutely sure of without a DNA test.

Meanwhile, back at the seagrass bed, somewhere in the world, the waiting is almost over for the male seahorse. His belly has swollen like a slowly inflating balloon and he is now ripe and ready to pop. For three long weeks he has barely moved from this spot, his isolation briefly relieved by visits from his partner. Every morning, throughout his pregnancy, they danced together just like the day they first met. Seahorse couples see each other only when they flirt and dance and mate. The rest of the time they spend apart, and as night falls and labor begins, the male is alone.

Three hundred miniature seahorses have wriggled themselves free from the pouch wall and energetically swim about, tickling their dad from the inside. Soon, he starts up with the same tail pumping action that successfully wooed his mate three weeks ago. Back then he was showing off how good he would be when this crucial birthing moment arrived. Now his pouch convulses in intensifying contractions. For some lucky fathers, delivery lasts a few minutes. For others the struggle can drag on for two or three days. In between heaves of exertion, the male stops and seems to catch his breath from the hard work of giving birth, his gills puffing and panting. In 1867, Samuel Lockwood described a pregnant male seahorse seeking some assistance from a seashell. "The winkle afforded real help in the labor of extruding the young," he explained. The seahorse "would draw itself downward and against the shell, thus running the pouch upward, and in this simple, yet effective way, expelled the fry."

And so, eventually, in multiple ejaculations, clouds of transparent specks like a swarm of apostrophes are launched into the sea, a herd of miniature seahorses with huge snouts too big for their spindly bodies but with all the necessary seahorse features already in place. The brand-new foals, each one the size of a flea, swim upward, inflating their swim bladders with a gulp of fresh air before drifting away to begin life with no more help from father or mother. They will settle down in different seagrass patches and after six months of feeding and growing, they will— all going well—find a partner and start a family of their own.

As for the proud fathers, their work is never finished. As soon as the arduous birth is over, the female returns and their courtship ritual resumes. The male may already be pregnant again by the next day, a tiresome life indeed but one that maximizes the output of offspring, which is, ultimately, all that really counts.

Now that we have answers to the questions, "What are seahorses?" and "Why do they look the way they do?" we are left with one last puzzle to solve: Where did they come from? This is a matter that for a long time was only wistfully pondered and not dealt with head-on. Scientists and naturalists preoccupied themselves instead with the simpler questions, "Where do they live?" and, more importantly for some, "How many different types can we find?"

In 1758, a Swedish naturalist, Carl Linnaeus[18] was the first person to give a seahorse a proper, scientific name. One of the most influential scientists of his time, Linnaeus wrote a pioneering text, *Systema Naturae*, in which he unveiled a revolutionary new way of looking at the natural world. Mankind has long been obsessed with trying to understand nature by dividing it up into well-defined units and giving everything an individual name, a practice of a branch of science known as taxonomy. But for a long time we got ourselves into a tangled confusion. Until Linnaeus came along, animals and plants were frequently described with dreadful mouthfuls. *Physalis amno ramosissime ramis angulosis glabris foliis dentoserratis* is a plant known nowadays as the Cape gooseberry or Chinese lantern. Linnaeus shone a much-needed beam of clarity onto this convoluted web, insisting that such long-winded names should be cut down to just two, thus the Cape gooseberry is now simply *Physalis angulata*. After initially focusing on plants, the tenth edition of *Systema Naturae* was the first time Linnaeus applied his classification system to animals, including *Hippocampus hippocampus*, the short-snouted seahorse.[19]

Linnaeus split the world into three great kingdoms: one for all the animals, one for the plants, and a third, oddly, for minerals. Although we don't stick to the same three kingdoms today—scientists have since discovered more kingdoms than Linnaeus could have counted on all his fingers and maybe even his toes—it is the way he subdivided the living kingdoms that we still use. According to Linnaean taxonomy, kingdoms are divided in turn into smaller groups called phyla (singular phylum), then again into classes, then into orders, then families, genera (singular genus), and finally into the smallest unit: species, defined as individuals that can interbreed and produce fertile offspring (horses and donkeys can breed but their progeny are sterile, so they count as separate species). Once you arrive at species, each one is given a double-barreled name in Latin, a politically neutral choice of language since by the eighteenth century Latin no longer belonged to any particular nation. The first name denotes the genus, like a surname, and the second is the species name, or forename. Thus humans, *Homo sapiens*, belong to the genus *Homo*, and we are the single species *sapiens*. *Hippocampus hippocampus* belongs to the genus *Hippocampus* and is the species of the same name.

Linnaeus's new system of nomenclature gripped eighteenth-century Europe and unleashed a watershed of naturalists and explorers who often referred to themselves as Linnaeus's disciples.[20] They packed their collecting kits, grabbed a map of the expanding empires, and set off with the single purpose of cataloguing as many living things as they could trap, net, pluck, or dig up. It was an era of unparalleled exploration and discovery, giving Europeans a new sense of global consciousness. Naturalists were assisted by free passage on overseas trading routes, especially the Swedish East India Company,[21] and they returned home laden with crates and pickling jars bursting with new species from faraway lands. Among the menagerie brought back by this new breed of international explorer were dozens of new seahorses.

The most important seahorse hunter was Pieter Bleeker, a Victorian doctor from the Netherlands. As a child, Bleeker longed for nothing more than to become a zoologist and to occupy himself counting and cataloguing wild creatures, but it was a career that his background did not easily permit. His father was a working-class sailmaker from a small town north of Amsterdam called Zaandam, and he was forced to leave school at age twelve to train as a pharmacist. Determined that he was destined for greater things, Bleeker saved up to pay his way through medical school and on qualifying, he tried, in vain, to get a job as a professional scientist. Eventually, in 1841, still at the age of only twenty-two, he signed up as a medical officer with the East Indian Army, where he hoped travel to the colonies would bring him closer to the natural diversity he had come to adore. And his dream did, finally, come true. Living in Batavia, now the Indonesian capital Jakarta,[22] Bleeker visited local fish markets and reveled in the sights and smells of so much exotic sea life. Right away, he realized that many of the obscure forms lying out for sale, their iridescent scales glistening in the tropical sun, must have been new to science and it was up to him to catalogue and name them. He set himself the ambitious task of writing a fourteen-volume *Atlas Ichthyologique des Indes Orientale Néérlandaises*, which he was still working on thirty-three years later when he died. He bought specimens from markets, preserving them in the local arrack rum, and recruited fishermen across the archipelago to hunt down unusual watery beasts. Despite being virtually cut off from mainstream scientific literature and with many other distractions, including studies in volcanism, ethnology, politics, and economics—not to mention his job as a doctor—he collected twelve thousand fish.[23] Bleeker identified five hundred new fish genera and nearly two thousand new species, many of which were beautifully illustrated in his *Atlas*. Among them were eight species of seahorse.[24]

In all, more than thirty people, most of them eighteenth-century naturalists, have named over a hundred different seahorses, described as "a century of sea-horses" in Gilbert Whitley and Joyce Allan's 1958 book *The Sea-Horse and Its Relatives*.

And these seahorses are found all across the globe. From San Francisco Bay and Baja California to the coast of Peru live seahorses dressed in suits of fine white scribbles. The waters from Nova Scotia to the Gulf of Mexico and as far south as Argentina are home to a cluster of seahorses, some with slender long snouts, some with pale saddles drawn across their backs, others darker and covered in tiny pinpricks like the night sky. The seahorses of northern Europe look truly horselike, with long spiky manes; and across the Mediterranean, along the Suez Canal and into the Red Sea, live smooth pale seahorses embellished with no patterns in particular. Seahorses live along thousands of miles of African coastline from the shores of the Congo Basin in the west, around the southern cape to the reefs of Mozambique, Kenya, and Tanzania. They live on sandy shores of Middle Eastern deserts and among crepuscular roots of mangrove forests of the Ganges Delta. To the east, there are seahorses in India and Sri Lanka, China and Japan. In Southeast Asia and islands of the Pacific lurk seahorses covered in long prickles like porcupines or striped like zebras, pink and knobbly or alight in a blaze of red flames. And Australia is blessed with a shroud of seahorses: Some are tubby, some are slender, some are squat and seem to shrug their shoulders. There are seahorses decorated in splashes of white paint, seahorses with glorious tails striped in black and gold, and some that look like nothing more than a clump of brown-green moss.

There is, however, one slight hitch. Many of these species were not meant to be. Of the hundred seahorses listed by Whitley and Allan, it turns out that a lot of them are synonyms of other species that had already been named as something else. The reasons for these duplications are twofold. First, seahorses are notoriously and frustratingly difficult to identify. For seahorses, color and body decorations aren't fixed like a polar bear's white fur coat, a tiger's stripes, or the fan of eyespots on a peacock's tail. Instead, they have at their disposal a wardrobe of apparel and, like chameleons, can change the color of their skin to suit every possible mood and for any situation they might find themselves in. Whether they want to blend with their surroundings or display their emotions, seahorses are always dressed for the occasion. Other fishes are much easier to identify based on their colors and patterns. Take the reticulated butterflyfish dressed in a gown of midnight black sprinkled in pale honeycomb flecks or the emperor angelfish that swims through its underwater kingdom wearing stripes of deep-sea sapphire and canary yellow, its eyes hidden behind a black mask edged in cerulean blue. And how about the three-spot angelfish, who sports a cheerful yellow coat decorated—of course—with three black spots, along with a smudge of blue lipstick as if it had been kissing the sky? There is not such an easy paint-by-numbers method for identifying seahorses. A few of them do have characteristic colorations, like the vivid stripy tails of tiger tail seahorses and Bleeker's lemur-tail seahorse. But even these can choose to swap colors and cover up their conspicuous tails if they want to. In 2006, a neon orange seahorse was found off England's southern coast.[25] Such a flamboyant creature had never been seen before, surely making this a new species. But no, it was in fact a well-known species, the long-snouted seahorse, which had changed color perhaps to blend with a sunken orange mooring buoy lying on the seabed.

Without color and texture to identify seahorses, taxonomists must rely instead on detailed anatomical measurements and there are only a limited range of characters to go on. They can count the minute bony rings that encircle a seahorse's body and tail, the number of transparent rays that prop up each fin, and they can calculate various length ratios, snout to head, tail to body. Even these apparently discrete dimensions tend to blend together at the edges. Fisher's seahorses have somewhere between thirty-six and thirty-nine tail rings, while Réunion seahorses have between thirty-four and thirty-eight.

The second major problem in deciding on seahorse species—or any species for that matter—is the lack of a centralized record of species and their names. Linnaeus' *Systema Naturae* was the last time anyone attempted to write a book listing together in one place the names of all the known animals and plants. Nowadays when scientists think they have good reason to, they publish descriptions of their new species in one of the many thousand peer-reviewed journals and deposit a "type" specimen in a museum somewhere. Specialists will keep an eye on their field of study, but it is virtually impossible to monitor everything that is going on. Several times, experts have attempted to draw together the sprawling world of seahorse taxonomy, but usually narrowed their focus to a certain region, the West Atlantic, say, or the Eastern Pacific. It wasn't until 2004 that researchers from Project Seahorse at the University of British Columbia decided it was time to sort through the jumble of seahorse taxonomy once and for all. Ph.D. student Sara Lourie took on the mission of sifting through more than a hundred putative species. She hunted down the original descriptions, visited museums to scrutinize the type specimens—some of them missing or damaged, some clearly not what they were supposed to be—and sat down to pick out all the duplicates and paint a global seahorse picture. In the end, she argued, there are thirty-three species of seahorse; at least there are thirty-three that are difficult to contest.[26] Since then, six new species have been discovered and described, adding up to a current grand total of thirty-nine.[27] All thirty-nine seahorses are united within the same seahorse genus, *Hippocampus*, and come from the same seahorse mold: Each one has a similar down-turned tubular snout (some more slender than others), pivoting eyes, a round belly (some more rotund than others), and curling prehensile tail.[28]

Even so, disputes continue and opinions are still divided between specialists who like to throw caution to the wind and split the seahorses into many more species, and those who more conservatively prefer to lump them together. But at the end of the day, does it matter if we know for certain that there are precisely thirty-nine or one hundred or five thousand species of seahorses living in the oceans? Well, yes, in many ways it does. When Linnaeus's "disciples" set out to catalogue wild fauna and flora, they weren't especially interested in studying the places where these new species were found; it was mainly a case of teasing apart the tangled threads of nature to impose a sense of order by sticking labels on the things they found. Since then the science of ecology has emerged, introducing a more holistic way of viewing nature, trying to comprehend how those tangled ecosystems work by following and studying connections between all living and nonliving things.[29] Every ecosystem, each coral reef, mangrove forest, pine forest, and savanna, is made up of gatherings of species, the fundamental units of ecology. Until we know which species are there and how each one lives, ecosystems remain impenetrable. Without the definition of species, we can't begin to understand how healthy ecosystems function or how our human meddling is unraveling them. There are undoubtedly more seahorse species waiting to be discovered and the picture of seahorse taxonomy is not set in stone, especially when we start looking at the seahorse's world from a new perspective that Linnaeus and Bleeker would never have dreamed of.

Since the discovery of DNA's molecular structure and with advances in genetic techniques, modern-day taxonomists no longer have to make decisions based purely on the way species look. Now they can peep inside cells and decipher a code of letters that distinguishes each living thing from all the others. We all carry a unique arrangement of DNA. Every person, giraffe, mosquito, oak tree, and seahorse is different at a fundamental molecular level, providing a powerful tool for deciding who is who and what is what. Surely this means the problems of seahorse identification and troublesome taxonomy are over? Extract some DNA from each potential seahorse species, look at how much each one differs, and there you have it: a definitive seahorse family tree. But of course, as you might have already guessed, it's not as simple as all that. There are a series of pitfalls and obstacles scattered on our path to taxonomic enlightenment. One of the biggest problems is deciding where, in genetic terms, to cut off one species from the next.

Imagine you are holding a length of silk ribbon dyed in all the colors of the rainbow; each color represents a different species. But where exactly does one species end and another begin? We might hope to see a block of orange in between a block of red and another of yellow. But instead what we have is a continuous spectrum of colors merging into each other; a band of scarlet in your left hand blends into deep orange, then lemon yellow; in front of you is a leafy green, blending into blue, then indigo, and finally in your right hand a deep royal violet. There is no magic number for how dissimilar the DNA of two individuals has to be—how different the colors on our ribbon—for them to belong to different species. We surely hope to see less difference within one species than between species, but that doesn't always happen. Sometimes, one species might sit well within the boundaries of orange, touching neither red nor yellow, while another might span all the way from mauve to dark green, through all the blues in between. Fortunately, this is not the case for our thirty-nine seahorses—at least not for all of them.

As well as using classical taxonomy in her global study of seahorse species, Sara Lourie also took a molecular approach. She wanted to find out if the species she identified based on their appearance, so-called morphotypes, would stand up to genetic scrutiny. Taking several hundred clippings from the fins of seahorses caught in shrimp nets in Vietnam, Lourie extracted from each one a stretch of DNA called cytochrome b.[30] Looking at the variation in cytochrome b in all her seahorses, Lourie must have been very pleased when she saw a perfect match between her seven morphotypes and the species divisions revealed by the messages hidden in their DNA. Seahorses belonging to the same morphotype had cytochrome b genes that were over ninety-nine percent similar. And when she hopped between morphotypes, the similarity in DNA codes broke down.[31] Lourie had a neat story of genetics supporting what she thought she already knew about species identity.

However, a few years later, some of Lourie's coauthors looked at cytochrome b again, this time in seahorses from various locations around the world. Now the story looked less clear-cut. While many of the morphotypes did fit in with the genetic patterns, the researchers stumbled on a few problems. They found a wide genetic gulf between seahorses from different locations thought to be the same species; three-spot seahorses from the Philippines were genetically distinct from three-spots in other countries; lined seahorses from the Caribbean had more in common with European short-snouted seahorses than with lined seahorses from Brazil. In other cases, there were seahorses that were split where perhaps they should be lumped; yellow seahorses and Cape seahorses were a bit too close for comfort; the same goes for slender seahorses and West African seahorses, which were difficult to tell apart despite living on opposite sides of the Atlantic.[32] Genetics, it seems, has helped firm up the perimeters of some seahorse species and cast doubt on others, leaving enough room for both the steadfast "lumpers" and the dedicated "splitters" to keep arguing their cases. But can genetics illuminate other parts of the seahorses' secret domain and help us answer the question of where they came from?

Modern genetics and more traditional seahorse studies do at least agree on one thing: All seahorses evolved from a single common ancestor, a pipefish. The question, though, is when did this evolution happen and where? Did it take place in the super-diverse "coral triangle" of the Indo-Pacific region (with the Philippines, Indonesia, and Papua New Guinea at its corners), where most of the world's shallow marine species, including the seahorses, now huddle? Or did seahorses begin life in the more impoverished Caribbean? As it turns out, these are the hardest questions of all to answer. Ideally, we could learn about the seahorses' distant past by searching through rocks for petrified remains of their predecessors. Unfortunately, fossil seahorses are even harder to track down than their living descendants. For a long time, seahorse fossils had been found only in five-million-year-old rocks from Italy. It was suspected that seahorses were in fact much older than that, because fossils had been found of close syngnathid relatives dating back at least fifty million years. Recently, however, another set of prehistoric seahorses has been unearthed in rocks in Slovenia.[33] These swam through much more ancient seas, around thirteen million years ago, giving a better guess as to how many birthdays the seahorse genus has celebrated. But we don't have to rely on fossils to look back into the past; once more, genetics can step in. By tracing genes that change slowly and steadily, we can scroll back further in time and find out which modern seahorse species are the most distinct from all the others and are therefore likely to be the oldest.

In 2004, a team of scientists from Rhodes University in South Africa, led by Peter Teske, confirmed that the oldest species of seahorse, the one that split off first from all the others, is the Bargibant's seahorse, or pygmy seahorse.[34] Today, they live across a wide swathe of the Indian and Pacific oceans, which doesn't help much in narrowing down where seahorses started off. The next two oldest species are the knobby seahorse and the big-belly seahorse, both of which live in Australia. So, could that be where they began? A few other pieces of evidence back up the theory of antipodean seahorse origins. Pygmy pipehorses are thought to be the intermediary step between pipefishes and seahorses. There are three genera of pipehorses; one lives in the Atlantic, one lives in various places around the Indo-Pacific, and the third is from Australia. These Australian pipehorses also happen to be the most seahorselike in appearance, hinting that they are the closest "nearly seahorses." Another inkling that seahorses were originally Australian comes from the fact that lots of seahorse species still live there today. According to Sara Lourie, a minimum of eleven seahorses live in Australian waters. Meanwhile, Rudie Kuiter, who specializes in Australian syngnathids, is sure the red continent is surrounded by at least seventeen more.[35] There have also been a few fossil seahorses found in Australia. Wakefield Wines in the Clare Valley of South Australia label their wine bottles with three entwined seahorses in honor of the rare seahorse fossils that were dug up during the excavation of the vineyard dam.[36] However, not everyone agrees with an Australian or even a wider Indo-Pacific birthplace for seahorses. Another group of researchers from the Institute of Zoology in London looked at similar genes but decided there was not enough evidence to pinpoint the seahorses' origin in either the Atlantic or Pacific Ocean.[37]

Not only does a question mark still hover over where the seahorses first began, but we also don't know for sure when they evolved. For many species, the speed at which the molecular clocks tick can be worked out using fossils. The age of the oldest known fossils of a species provides an estimate of when it split off from other species (although there is always the possibility that someone will find an even older fossil). By working backward in time and down through the family tree, the age of an entire lineage—like the seahorses—can be determined. But as we have already seen, the seahorse fossil record is, for some reason, eerily empty.[38] So instead of using fossils to calibrate molecular clocks, scientists have to rely on other hints to date crossroads on the family tree. Take for example two similar species, the slender seahorse and Pacific seahorse that live today in the Caribbean and Pacific, one on either side of Central America. We might assume they started off as a single species that was split and two subpopulations forced to part company when the Isthmus of Panama slammed shut around three million years ago, joining North to South America. That gives an earliest possible date when the two new species could have set off down their own evolutionary pathways. Working backward from the slender/Pacific seahorse split brings us to a rough estimate that seahorses are at least 16.5 million years old.

For argument's sake, let's say that seahorses did first evolve in Australia. That leaves us with one final question: How did the slow seahorses come to occupy such distant reaches of the vast oceans? Unlike most other fish, seahorses don't shed thousands or millions of eggs into the open water, sending off their progeny to drift for days or months at the mercy of ocean currents. Baby seahorses are born well developed and usually don't

move far from their birthplace before settling down and more
or less staying put. Could adult seahorses instead have cruised
around the oceans hanging on to something else?

Seahorses are perfect candidates for long-distance, panoce-
anic hitchhiking. Instead of swimming or drifting through the
water column, they can use their strong tails to hold on tight to
rafts of floating seaweed that provide some protection from the
elements, a place to hide from predators, and come laden with
supplies of food so passengers need not go hungry. And all it
takes is a single intrepid pregnant male, like a boat full of New
World settlers, to initiate a new population on a distant shore.

Several genetic studies support theories that seahorses sail
across ocean basins.[39] Within separate populations studied in
Africa, India, and Southeast Asia, every seahorse shares the same
version of an ancestral gene inherited from one pioneering individual that probably disembarked from a seaweed cruise liner
and established the population in the first place. Also, if seahorses were swimming around the oceans under their own fin
power, we would expect to see a relationship between genetics
and geography; populations closer together would have more in
common genetically than those farther apart. But that is not
what researchers have found. Instead, there are random patterns of genetic similarity across wide areas of ocean, reinforcing
the idea that seahorses do indeed hitch long-distance rides on
clumps of oceangoing tumbleweed; once a seahorse hooks on
to a tangle of seaweed, it can cling on and survive for however
long it takes to reach a suitable patch of coastline—that could
be nearby or it could be thousands of miles away.[40]

In recent years, modern science has taught us much about the
creatures roaming through the ocean depths; we know how
long they live, what they eat, who their ancestors were, and how

it was that they came to colonize the sea. Even the shy and mysterious seahorse has begun to give up its secrets to genetic probes and keen-eyed scientists. We are now closer than ever before to understanding where they fit in the animal kingdom and how and why they abandoned so many aspects of a more conventional fish way of life. But has all this knowledge planed away the beauty and intrigue that surrounds them? Maybe not. There is still a lot we don't know; there are still species to be discovered and more questions to be answered. And maybe the science of seahorses can never truly answer the eternal question of why we find them so utterly captivating and so different from all the other creatures in the sea.

Science has also unveiled just how much of our human history we have shared with fish and other aquatic creatures. When hunter-gatherers first began migrating out of Africa around 125,000 years ago, it is thought they harvested giant clams from the coral reefs of the Red Sea, possibly driving the largest, most nutritious species into catastrophic demise.[41] In the Congo Basin of West Africa, archaeologists dated the remains of bone harpoons and piles of fish remains to around 90,000 years ago.[42] With such ancient origins, mankind's long-lasting relationship with the world's fishes has been based, for the most part, on utility. While aquarium keepers and collectors have not overlooked the aesthetic appeal of many marine and freshwater species, the most common use for fishes has been to fill our bellies. But something about the seahorses' bizarre yet irresistible essence has led people to treat them differently from other fish, finding many other uses for them in our human world.

Chapter Three

A SEAHORSE CURE

There was once a great Chinese emperor named Shen Nong. If we could travel back in time five thousand years we might spot him, his skin tinged a sickly shade of green, as he clambered down a set of rickety ladders from the highest peak of the Daba mountain range in the Hubei Province of central China. Inside a cloth sack slung across his back are bundles of flowers plucked from the mountainside, not for their colorful blooms or intoxicating scent but for their healing powers. The mountain was named Shennongjia, or Shen Nong's Ladder, a region still renowned for its rich diversity of medicinal plants and, some say, home to the elusive Chinese yeti.[1] Legend has it that Shen

✦

Above: Woodcut of two Mediterranean seahorses from Pietro Matthioli's
Commentaries on the Materia Medica of Discorides, 1565.

Credit: Pietro Andrea Matthioli, 1565

Nong dedicated much of his life to traveling around China, scouring the landscape for natural remedies to treat the illnesses that plagued his country. He used himself as a guinea pig, tasting all the plants, animals, fungi, and minerals he found, risking daily poisoning and gaining an unnatural pallor. It is also said that Shen Nong came up with the idea of sprinkling leaves of the *Camellia sinensis* plant into boiling water to make the world's first cup of green tea, an antidote for the substances he tested that turned out to be poisons, not cures. Also known as the Divine Farmer or Emperor of the Five Grains, Shen Nong is celebrated as a great hero of Chinese antiquity, in part because he taught the Chinese people how to cultivate crops. But he is perhaps best remembered as the forefather of Chinese herbal medicine, a branch of medicine that remains popular today among millions of people.[2]

Shen Nong's name appears in the title of the first-ever book of Chinese medicines, the *Shen Nong Ben Cao Jing*, otherwise known as Shen Nong's herbal classic. Original copies of the book have never been found and no one believes that Shen Nong in fact had anything to do with writing it. Almost certainly it was compiled by a team of scholars from the Han Dynasty in the first century BC, who dedicated their work to the character they regarded as the founder of Chinese herbal medicine.

Modern reconstructions of the *Shen Nong Ben Cao Jing* reveal that it contained descriptions of 365 plants, animals, fungi, and minerals, all referred to as "herbs"—one for each day of the year.[3] Most of them were plants or parts of plants, including such things as rhubarb root, forsythia fruit, bugbane rhizome, Chinese honey locust fruit, marsh orchids, asparagus tuber, licorice root, mimosa tree bark, and Japanese teasel root.[4] The prospect of eating some of the animal ingredients would make even the toughest stomachs wobble, including earthworms, centipedes, snake skin, hornets' nests, and the eggs of a praying mantis laid on a mulberry leaf. Seahorses, however, did not feature in Shen Nong's menagerie.

As China descended through successive dynasties, dozens of reworked versions of the *Shen Nong Ben Cao Jing* were composed and the list of herbal medicines grew and grew. Some medicines were brought back from far-off lands as traveling Chinese physicians swapped secrets with herbalists from India, Korea, Japan, and the Middle East. One of the most influential compendiums of Chinese herbal medicines was the extravagant fifty-two-volume *Ben Cao Gang Mu* written during the Ming Dynasty in the sixteenth century. It was the lifework of a famous Chinese doctor, Li Shi Zhen, who dedicated himself to undoing what he saw as the many errors and faults that peppered earlier texts. Like Shen Nong before him, he traveled around China gathering information about herbal remedies, his efforts culminating in the description of nearly two thousand medicines. One of the medicines he listed in his text was the seahorse.

> Hai ma [seahorse]: sweet taste, warm property. Supplements kidney. Strengthens yang. Removes lumps and masses in the lower abdomen of women. Treats furuncles and toxic swellings, dystocia, and pain due to malfunctions in blood and vital energy.[5]

Four hundred years later, belief in the healing powers of dead seahorses lives on. Roasted and pulverized into a fine powder, then stirred into a pungent soupy tea or steeped whole in yellow rice wine, seahorses are prescribed to treat an astonishing variety of human conditions. Among a prospering Chinese middle class, both in China and abroad, seahorses are a pricey medicine used to help mend broken bones, to calm asthmatic lungs, to gain control of a leaky bladder, and to soothe inflamed boils, pustules, and ulcers. Seahorse powder and tablets are administered to treat throat infections, to reduce high cholesterol, to speed the healing of open wounds, and for people suffering from kidney and liver disease. Seahorses are even considered by many to be a natural remedy to perk up a man's flagging libido.[6]

A craze for seahorse medicines rages on not only in Chinese communities but also elsewhere in Asia and across the world. In traditional Japanese Kanpo medicine, seahorses are used as an aphrodisiac. In Central and South America, a dose of seahorse is used to treat asthma. Indonesian Jamu medicine largely remains the secretive domain of male tribesmen, but it is known that they use seahorses for treating impotence, loss of memory, and rheumatism. And in Tamil Nadu in India, ground dried seahorse mixed with honey is given to children with whooping cough.

Several of the seahorses' close relatives have also found themselves on the list of ingredients used as herbal medicines. Dried pipefish is a common Chinese cure, deemed to pack twice the therapeutic punch of seahorses and valued most highly when their tails are straight and unbroken. An unusual treatment for bronchitis is seamoth—not, in fact, a flying insect but a type of flattened fish that lies on the seafloor with its winglike pectoral fins stretched out sideways.[7] Medicines are also made from pipehorses, the odd creatures that look like a cross between a pipefish and a seahorse. Of all the members of the Syngnathidae family, only the seadragons have escaped the attention of Chinese pharmacies, hiding as they do in small numbers along the southern coast of Australia, away from the direct gaze of the Asian medicine trade.

In Chinese and other Asian cultures, there is no clear distinction between ingredients used as medicines and those eaten on a daily basis to promote general well-being. The sea provides many "health foods," like the gloopy noodles of cartilage extracted from dismembered sharks' fins and made into soup. Sea cucumbers, or bêch-de-mer, are tubular relatives of starfish that lie on the seafloor like giants' fingers and are cooked up into another amorous potion thought to boost masculine prowess. Seahorses also feature on restaurant menus and food market stalls around Asia; you can have boiled seahorses added to a bowl of soup or garnishing a stir-fry, or nibble on a row skewered side by side on a barbequed seahorse kebab. Vietnamese fishermen make an energy-giving tipple for long nights at sea by dropping small seahorses into a bottle of whisky.[8] And in Peninsular Malaysia, fishermen add ground-up seahorses to their coffee and deep-fry them to make a crunchy snack.

As medicines blend into health foods, so there is also a hazy crossover between health foods and more fantastical notions of the effects seahorse preparations can have. Pregnant women across Asia wear a dried seahorse on a cord around their necks and, at the onset of labor, use it to quickly brew a cup of tea that will promote a trouble-free birth. If there is no time to put the kettle on, mothers-to-be can simply hold on to a dead seahorse to ensure a smooth delivery and bestow good luck on their baby. In Malaysia and the Philippines, fishing communities hang dried seahorses in doorways or around the necks of children as a talisman to dispel evil spirits. In Indonesia and Mexico, lucky seahorses are used to protect money, bring prosperity to businesses, and grant fishermen good fortune at sea. Biologists studying the East African seahorse trade met a fisherman on the island of Zanzibar who had been told by a folk healer to use seahorses to boost his poor catches. He would take a burned seahorse, mix its ashes with a medley of plants, and then sprinkle the concoction over the fishing gear to banish bad spells and attract fish into the nets.[9] And humans are not the only animal that can apparently benefit from seahorses. In 1834, surgeon and naturalist George Bennett arrived in the Dutch colony of Batavia, following his wanderings around Australia in search of the baffling duck-billed platypus. During his stay a local presented him with a dried seahorse wrapped carefully in paper. The Malays called it *ecan kudu,* or "horse-fish," Bennett reported. He was instructed to drop the dead seahorse, whole and otherwise unprepared, into a horse's water trough where it would act as an excellent equine tonic.[10]

The most dubious claim for the supernatural abilities of sea-horses has been that by eating them, people can tap into a fishy way of life. Bernard E. Read, an English pharmacologist working in China, wrote in 1939 of an earlier Chinese physician called Pao P'u Tzu who claimed "seahorses, with red speckled spiders and P'ing Yi's water-fairy pills when eaten will enable one to live underwater." Read, however, skeptically adds, "At present time the water-fairy pills are without potency."[11]

Despite having a long history of herbal medicines, it is not obvious exactly why the Chinese started using seahorses as a natural remedy. What we do know is that they were probably not the first people to do so. Long before the idea caught on in Asia, ancient Europeans plucked seahorses from the shores of the Mediterranean and made them into a range of medical concoctions. During the reign of the Roman Empire, supernatural imaginings of Hippocampus spilled over into real life as people began to believe that seahorses held the power to heal—and also to kill.

Much like ancient Chinese medicine, early European pharmacology revolved for centuries around a key textbook that listed powerful natural ingredients, instructing on their use and disuse, how to harvest them from the wild and how to prepare them. But unlike the surviving fragments of the original *Shen Nong Ben Cao Jing*, a complete copy of the seminal European text still exists, entitled *De Materia Medica libri cinque* ("Concerning medicinal matters in five volumes"). The oldest remaining original copy dates from 512 AD and was a gift to Anicia Juliana, a Byzantine princess from Constantinople. For centuries her copy of *De Materia Medica* was well used and passed among many scholars and physicians. In the margins of a thousand ancient pages, notes and amendments are scribbled in Arabic, Turkish, Hebrew, and French, a reflection of the many nationalities that over the years made Constantinople their home. Eventually, in 1567, Emperor Maximilian II bought Juliana's book and donated it to the national library in Vienna, where it remains today.[12]

The author of the book was a man named Pandianus Dioscorides who lived in the first century AD. He was born in Anazarbus, a small town on a bend of the River Pyramus in the Roman province of Cilicia, now part of modern-day Turkey. After studying medicine in Tarsus and Alexandria, Dioscorides traveled widely in Asia Minor as a surgeon with the armies of Emperor Nero. As he went, he sought out local doctors and herbalists, gathering information and observations of herbal treatments. Although other European scholars before him had written books of herbal medicines, the five volumes written by Dioscorides were the most complete and thorough to date. He had an urge, like Li Shi Zhen, to fill in the gaps and correct the mistakes of earlier works. On completion, Dioscorides' *De Materia Medica* met with instant acclaim across the Roman world and, even though the author himself remained obscure, his book became a household name. It was translated into many languages and regarded as the authoritative reference book on medicinal ingredients, remaining in popular circulation throughout Europe for the next fifteen hundred years.[13]

In his book, Dioscorides describes how a mixture of burned seahorse and goose fat (axungia) produces a paste to smear on balding scalps in the hope that it will coax the return of a full head of hair:

> Hippocampus is a little living creature of the sea, which, being burned and the ashes thereof taken either in Axungia, or liquid pitch, or unguentum Amarcinum and anointed on, doth fill up the Alopecia with haire.

Several other Roman writers, inspired perhaps by Dioscorides or even the other way around, also wrote of the restorative powers of seahorses. The Roman natural historian Pliny the Elder gave a seahorse recipe to treat various skin complaints:

> Lichens and leprous spots are removed by applying the fat of a sea-calf, ashes of mæna in combination with three oboli of honey, liver of the pastinaca boiled in oil, or ashes of the dolphin or hippocampus mixed with water.[14]

Pliny also recommends seahorse as an aphrodisiac, as a cure for urinary incontinence, and grilled to treat "pains in the side." Seahorse killed in rose oil was intended to fend off chills and fever. And seahorse "taken in drink" was also recommended as a cure for bites from the sea hare, a harmless type of sea slug that, when disturbed, squirts a jet of purple ink in its tormentor's face. Such colorful self-defense may explain why it was thought these squidgy sea creatures had magical powers and were poisonous to touch.

Another account of the curative capacity of seahorses comes from Claudius Aelianus, also known simply as Aelian, a Roman writer and teacher of rhetoric who lived in Italy at the turn of the third century. He wrote two major works, *Historical Miscellany* and *On the Nature of Animals*, both entertaining collections of facts, fables, and anecdotes about local customs and natural wonders, with a thin thread of moral allegory woven throughout. Some of Aelian's stories were clearly absurd inventions, and he openly admitted that they were merely stories he had picked up from others that he didn't believe to be true. Among the many peculiar stories he told, possibly to amuse and not inform, was the outrageous self-castration of a beaver. In Roman times and beyond, beaver testicles were widely valued as a medicine to treat nervous conditions. Aelian wrote about a wise beaver who, when pursued by hunters, reached down and chewed off the treasured items he knew the hunters were after, handing them over "by way of ransom." If ever the unfortunate beaver was cornered a second time, he would stand up and reveal to his pursuers that there was nothing left of interest. (The lesson to be learned is perhaps that we might benefit from vanquishing vice from our lives.)[15] Aelian did, however, take a somewhat more serious and personal interest in matters of the ocean, retelling many accounts he gleaned from Mediterranean fishermen. One tale tells of an old fisherman whose sons were bitten by a mad dog and, while everyone else was panicking and praying for divine assistance, he calmly reached for a couple of seahorses he had caught in his fishing nets earlier that day. As Aelian wrote:

[He] washed out the stomachs of the Sea-horses, some of which he roasted and gave to the boys to apply, while others he pounded into a mixture of vinegar and honey, and

then smeared on the wounds made by the bite, and so over-
came the boys' madness by that longing for water which
the Sea-horses engendered in them. And in this way he
cured his sons, though it took time.

Aelian also tells a second, more gruesome account of the effect
of seahorses:

Those who are expert at fishing say that if one boils and
dissolves in wine the stomach of the Sea-horse and gives it
to someone to drink, the wine becomes a poison abnor-
mal in comparison with others.

He goes on to describe the horrifying repercussions of seahorse
poisoning:

For the man who has tasted it is first of all seized with a
most violent retching; next he is racked with a dry cough
but brings up nothing at all yet his upper stomach is en-
larged and swells, while hot streams mount to his head and
phlegm descends from his nose, emitting a fishy odour; his
eyes turn bloodshot and fiery and the lids become puffy.[16]

If the victim does not vomit, his fate is certain death, Aelian
tells us. Those who do manage to evacuate the noxious wine will
pull through only to become infused with an irrepressible desire for water; they find themselves addicted to the soothing
sounds of babbling rivers and streams and spend the rest of their
lives lolling on beaches and swimming in the sea. The sugges-
tion that seahorses could instill a longing for water might explain their use as an antidote to rabies since one symptom of

the disease is hydrophobia or a terrifying fear of water. A hint that Aelian's chronicle of seahorse poisonings had some rational basis comes from his admission that perhaps the seahorse itself was not the perpetrator, but instead a type of bitter seaweed that they eat. Although seahorses are not known to be avid vegetarians munching through large amounts of seaweed, some fish eat types of single-celled algae that contain poisons and can accumulate to lethal doses.[17] Perhaps this third-century writer was not so far off the mark.

The compass point of early European medicine spun around so frequently—there was even a time when people believed demons were the root cause of all disease—it is remarkable that Dioscorides' book was still in demand a millennium and a half after it was written. During that time, the popularity of seahorses quietly persisted. A 1753 issue of *The Gentleman's Magazine* reported that Italian high-society ladies were using seahorse potions as a means of "increasing their milk."[18] However, as Europe entered the nineteenth century, herbal medicines began to be nudged aside. With the growing acceptance that germs cause most diseases, traditional medicines were eventually abandoned, more or less, in favor of new synthetic drugs emerging from the fast-moving science of chemistry. A handful of natural remedies still live on in the modern western world. Some find informal uses like dock leaves to soothe a stinging nettle rash; others, like the purple flowers of foxgloves, have been interrogated by science and found to contain powerful molecules, giving them a brand-new role in the armory of modern drugs. But even with the recent resurgence of interest in herbal remedies and alternative therapies, seahorse cures for baldness or an antidote for rabies have not survived the test of time. No longer can the dried bodies of these odd-looking fish be found hiding on the shelves of western pharmacies or locked away in bathroom cabinets.

As well as failing to maintain their influence into the twentieth century, there is another major difference that sets apart the early medicinal textbooks from East and West: Dioscorides steered clear of signing up with any single school of medical thought. Instead, his work was based entirely on what he thought were indisputable observations of the effect each medicine had on the human body. It didn't matter what you believed caused a stomachache as long as the remedy worked effectively against it. He made no claims as to exactly how or why seahorses were a cure for baldness—that was simply what he, or folk healers he met during his travels, saw happening. Dioscorides' work presumably weathered the changing medical climate in ancient and medieval Europe because it fitted into virtually every medical system that came and went.

In contrast to the shifting medical beliefs in Europe, Chinese medicine has remained far more constant. Every single version of the Chinese herbal textbooks, from Shen Nong's onward, has been based on the same core philosophy that explains what causes illnesses and how good health can be restored.

Back in the sixteenth century, Li Shi Zhen described dried seahorses as being "warm" and "sweet," with the ability to strengthen "yang" and treat an excess of "yin." These words may be ancient, but they are neither outdated nor have they fallen out of use. Instead, they reveal part of a trelliswork of ideas on which Chinese herbalists still hang their understanding of how the human body gets sick and what can be done to heal it. According to the medical discipline known today as Traditional Chinese Medicine, often shortened to TCM, the secret to a healthy life is maintaining a harmonious balance within the body. And the very essence of that balance is yin and yang.

To many people, the words *yin* and *yang* conjure an image of a black-and-white disc; within it are one white and one black tadpole hugging neatly together, each one peeping out with a single round eye the contrasting color to its body. This Taoist symbol—much hijacked by new age beliefs—was designed to represent an ancient Chinese way of understanding how things in the universe work. Originally, and quite simply, *yin* referred to the shady side of a hill, and *yang*, the sunny side. Later, the two words came to represent any pair of polar phenomena with opposite but at the same time complementary characters: night and day, hot and cold, fast and slow, heaven and earth, and so on.

It is said that all things in the universe are made up of a combination of two aspects: a yin aspect, which is dark, cold, lower, inward-looking or resting, and a yang aspect, which is bright, warm, higher, outward-looking or moving. These two aspects are intimately related to each other according to a set of fundamental rules that apply to the universe as much as they do to the human body: Nothing can be purely yin or purely yang, merely that one or the other predominates; neither can exist without the other, thus something tall is tall only with something short to compare it with; yin and yang mutually control and inhibit each other; they exist in a state of constant flux and they transform into one another. We see all these things taking place with the passing of the seasons. When the cold weather of winter ebbs, the warmth of spring and summer flows until the heat diminishes and the season transforms once again into autumn, then winter.

Fundamentally, according to TCM, it is an internal imbalance of yin and yang that causes ill health. There are many different aspects of yin and yang within the human body and a multitude of ways that disease-causing imbalances can occur.

Some organs pertain to yin, some to yang, depending largely on where in the body they are found and the sort of functions they carry out. Organs higher in the body are yang; lower organs are yin. Organs linked to the outside of the body are yang while those found deep inside are yin. In a healthy person, the overall balance of yin and yang is maintained by the flow of an invisible and undetectable substance called qi or chi (pronounced "chee"). There is no English word or phrase that adequately captures the meaning of qi, but it is sometimes known as "vital energy." Everything in the universe, both organic and inorganic, is made up of qi, but it is not some sort of primeval matter, nor is it strictly a force of energy. Some of the body's qi is inherited at conception, some is breathed in from the air, and some is taken in with food and water. Once inside, qi flows through channels called meridian lines, which create a web of interconnections around the body. If the supply of qi is inadequate or its flow around the body is hindered or blocked, then yin and yang are likely to become imbalanced and illness will follow.

Diagnosis of poor health and disease in TCM focuses on interpreting a complex web of signs and signals that quietly whisper the underlying imbalance of a patient's yin or yang. During a lengthy consultation, a traditional Chinese doctor will examine a patient's appearance, noting the paleness or pinkness of their skin, peering closely at the color and texture of their tongue, listening to their voice, feeling for the rhythm and vigor of pulses on both wrists, and asking in-depth questions about the patient's moods, relationships, and general perspectives on life. All these are vital clues to the patient's internal condition, since outward symptoms are thought to be merely manifestations of what is going on inside.

Once a pattern of disharmony has been identified, the doctor prescribes treatment to bring the patient back to a balanced and healthy state. Flow of qi can be physically stimulated by the pricks of acupuncture needles, while herbal medicines take a chemical approach to restoring internal balances. Modern textbooks list more than ten thousand Chinese herbal medicines, each one classified according to two key properties—temperature and taste. Medicines can be hot, warm, cold, cool, or neutral, as well as sweet, sour, bitter, spicy, or salty. In various combinations, the temperature and taste of an herbal ingredient determine the influence it has on the balance of yin and yang in the body. Yin herbs are cold, sour, bitter, and salty and in general are used to treat yang conditions such as the excessive heat of a high fever. The hot, warm, sweet, and pungent herbs, including seahorse, belong to yang and help to redress an excess of yin or a deficiency of yang.[19]

Much of the therapeutic action of seahorses is thought to take place through the kidneys—yang organs that are connected to the outside of the body. In the eyes of western anatomists, the role of the kidneys is to filter blood and expel excess water and waste products from the body, such as salts and urea, in the form of urine. To a certain degree, traditional Chinese theory takes the same view, which is why seahorses are used to treat problems like urinary incontinence. But it doesn't stop there. Traditional Chinese physicians consider kidneys to be among the most important organs in the body. Qi is breathed in through the lungs and transported directly along a meridian line to the kidneys, which store it, transform it, and redistribute to other parts of the body. When the kidneys are weak, in particular if they are deficient in yang, they are unable to grasp on to qi from the lungs. Thus, by strengthening the kidneys, dried seahorse is used to treat breathing problems like wheezing and asthma.

Replenishing kidney yang is also the key to using seahorses as a treatment for impotency. The traditional Chinese view of sex focuses on an exchange of yin and yang. During intercourse, the woman gives yin to the man, who in turn gives yang to the woman. When a man doesn't have enough yang, his performance may be disappointing. Ancient medical texts also allude to the pairing of male and female seahorse "lovers" in medicinal preparations to enhance their potency for inflating male accomplishment. Seahorse couples are sold in markets and pharmacies across Asia, although often they are paired up based on size rather than sex.

An important aspect in which TCM differs from modern western medicine is that, in China, patients are treated holistically and individually. The aim of TCM is not to identify a disease and match it to a specific drug. Instead, TCM doctors skillfully compose a subtle fusion of herbs that will treat the unique pattern of disharmony presented by each patient. Therefore, despite their apparently potent effects on the human body, seahorses are not prescribed alone but always in a complex cocktail of ten or twenty other ingredients. A prescribed blend will contain a powerful chief ingredient that treats the main symptoms plus others to enhance the overall affect, some to treat accompanying symptoms, others to counteract any toxic side effects, and some that guide the treatment to specific parts of the body.

At the Chinese pharmacy, the prescribed ingredients are selected from rows of wooden drawers and glass storage jars lining the walls and mixed together in a bag for the patient to take home. The customary way to consume Chinese medicines is as a pungent tea that is steeped in boiling water for an hour and fills the house with a potent stench. The resulting infusion, known as a decoction, is drunk several times a day. As the patient's condition alters day by day, the doctor can intricately tweak the blend of herbs to achieve the desired restoration of internal balance.

The questions that must hover in many western minds are "Do Chinese herbal medicines work? Can a dose of dried seahorse really cure impotency, asthma, and bed-wetting? Are herbal remedies a real match for the diseases of the twenty-first century?" These are questions that may ultimately wriggle free from a simple yes or no answer. Perhaps we should listen to the cries of a billion people as they proclaim a resounding and unified "Yes." One third of all people living today are thought to rely on the knowledge of traditional medicines for much of their health care needs.[20] Such widespread and growing popularity is hard to ignore. Can all these people be wrong? Maybe not, but surely it is crucial that these herbal medicines on which so many people depend should be tested for their safety and efficacy.

Some claim that time is testament enough to demonstrate that traditional medicines work. The same herbal remedies have been used in China for hundreds, and some for thousands, of years.[21] In spite of its ancient and convoluted past, TCM has survived as a distinct discipline throughout the comings and goings of dozens of dynasties that grabbed their turn ruling the country. Apart from the exchange of some medicines with other cultures, Chinese medicine remained remarkably untouched by outside influences, often through policies of deliberate isolation. Even when the Renaissance began to take hold in Europe and modern science-based medicines emerged, only the small-scale efforts of missionaries brought a few western drugs into China. If Chinese herbal medicines didn't work, why, then, are they still used after all this time?

TCM is so persistent that even calculated attempts to wipe it out have failed.[22] When the Chinese Nationalist party, the Kuomintang and then the Communists seized power in China in the first half of the twentieth century, they both made efforts to eradicate traditional medical practices, seeing them as regressive ties shackling the nation to their feudal, undeveloped past. A prominent Chinese Marxist, T'an Chuang, notoriously referred to TCM as the "collected garbage of several thousand years."[23] Public health officials from the Kuomintang presented "A case for the abolishment of old medicine to thoroughly eliminate public health obstacles," which would have banned the advertisement of Chinese medicines, restricted its practice, and prevented the establishment of TCM schools. Following impassioned public outcry, the proposals were not implemented, but they firmly set the mood for the practice of TCM for decades to come.

It was 1929 when Mao Tse-tung, agreeing with the beliefs of the ruling Nationalist government, commanded his Red Army to "uproot all shamanic beliefs and superstitions."[24] But by the time he had taken charge of the country, he quickly realized that traditional medicines could not and would not be stamped out. In a famous U-turn in 1949 he declared that "Chinese medicine is a great treasure-house. We must make all efforts to uncover it and raise its standards." He had discovered there were too many people, and in particular too many poor people, in China who had no alternatives but to use cheap herbal medicines. These traditional remedies were widely available across the vast landmass of China while western drugs were expensive luxuries restricted to larger cities. Many people firmly believed—as many still do—that traditional medicine can effectively treat some conditions where western medicine fails. And so a new institutionalized form of traditional medicine was repackaged, relaunched, and heralded as the way forward.

One of Mao's central initiatives, as part of his Cultural Revolution in the 1960s, was to mobilize a national workforce of "barefoot doctors." He offered rudimentary medical training in both traditional and western disciplines to anyone who wanted it, no matter what their background or experience. They were sent out into rural China to provide cheap medical services to the masses and today there are still over a million of them living and practicing in areas where urban-trained physicians would not live and work.

The alliance of western medicine with traditional techniques was another aim of the new regime, with institutes established around the country to foster relationships and encourage exchange between the two. Mao also wanted to provide his own answer to the question "Does it work?" He demanded that the spotlight of scientific inquiry be pointed directly at traditional medicines, confident that they would stand up to the glare and unveil their value to the rest of the world. And, in many ways, he was right.

The main way that western science tests whether or not Traditional Chinese Medicines work is by rummaging through them for individual substances that are biologically active and can be isolated and proven to have clinical effects. And indeed scientists have found many powerful molecules and compounds this way, most of them from the plant world. Essentially, the reason any living thing goes to the trouble of making complex chemicals, often requiring lots of energy to do so, is for the purposes of attack or defense. In the animal kingdom, chemical weapons are rarely employed since simpler tactics like running or swimming—in pursuit or escape—are usually preferred. Warm-blooded mammals and birds are always ready to flee, and very few have evolved a poisonous bite or toxic skin. It is mainly among the cold-blooded animals that we find interesting chemicals, in the lethal vibrancy of poison arrow frogs, in a stonefish's deadly impersonation of a weedy rock, and in the summertime sting from a sat-upon wasp.

Throughout human history, plant-based remedies have dominated traditional pharmacopoeias ultimately because plants adopt a hard line in chemical defense. Compared to the itinerant animals, plants are rooted—quite literally—to the spot. Their tender green leaves spread out to collect the sun's energy and they offer a tempting and immovable feast for any passing herbivores. Plants evolve spines and prickles to ward off gnawing intruders, and many taint their vulnerable tissues with a variety of sophisticated compounds, ranging from bitter and bad-tasting to downright deadly. Without even knowing it, mankind has been exploiting these molecules and compounds for millennia by seeking out and eating plant extracts that, when taken in the right doses, will treat many human conditions. Humans are not the only ones with an instinct for the curative powers of plants; many other animals are good at it, too. When they are laid low with upset stomachs, chimpanzees and gorillas have been seen chewing the pithy stems of the ewuro shrub; the bitter juices have little nutritional benefit, but they do contain biologically active compounds that deter intestinal parasites. Starlings keep diseases at bay by lining their nests with leaves laced with antibacterial compounds, and capuchin monkeys rub crushed citrus fruit into their fur to help repel mosquitoes and heal skin infections.[25] It comes as no great surprise, then, that western scientists have reached the conclusion that many plant-based Chinese medicines do indeed work.

Mao was right when he had a hunch that Chinese herbal medicines would surrender a cure for one of the world's deadliest diseases. In 1967, driven by a desire to stop losing his soldiers to malaria, he called on Chinese military researchers to screen hundreds of traditional antimalarial medicines for any signs of a scientific basis for their efficacy. It wasn't until 1972 that researchers found what they were looking for, when they stumbled on a thousand-year-old recipe for a type of tea known as qinghaosu. A distillate of the tea yielded a molecule that proved to be the most powerful antimalarial drug in history. It clears the blood of stubborn malaria parasites that have evolved resistance to other drugs, and it is especially effective against the most dangerous form of the disease, caused by the parasite *Plasmodium falciparum*.[26]

The new drug was named artemisinin after the woody shrub it is extracted from, *Artemisia annua,* also known as sweet wormwood or sweet Annie.[27] Artemisinin should have been an overnight, global success and yet, for several decades, hardly anyone outside China had heard of it.[28] Amid the distrustful atmosphere of the Cold War, the Chinese were reluctant to tell anyone where their wonder drug came from and the rest of the world was skeptical. The structure of the artemisinin molecule, as published in Chinese academic journals, was completely different from all the other antimalarial drugs, leaving western scientists doubtful that it could work. And without access to the drug, they couldn't test it themselves. Eventually, in 1984, members of the U.S. military found *Artemisia annua* growing as a weed on the banks of the Potomac River in Washington, DC, and after further tests they finally agreed with the Chinese.[29] Artemisinin, thanks to its unusual and unstable structure, really does work. Chinese medicine has brought about a new, highly effective drug that is now being widely used in the fight against malaria, particularly in Africa and Southeast Asia. The only problem is that, if used incorrectly, there is the risk that artemisinin will have only a limited shelf life as a useful antimalarial. The World Health Organization has issued recommendations that artemisinin should be used only in combination with other drugs, so-called Artemisinin Combination Therapies or ACTs, to minimize the chances that malaria parasites will rapidly develop resistance to the new drug as they have done with many other malaria treatments.[30] The drawback is that ACTs are complex and expensive to produce. A 2008 study revealed that one-third of antimalarial drugs on sale in Africa were "monotherapies" made up of artemisinin alone.

And what about seahorses? In many ways, seahorses are quite plantlike. They aren't fast-moving, which might make them one of the few animals that is chemically equipped to deal with uninvited guests. But so far scientists have failed to isolate any novel chemical deterrents from seahorses, no artemisinin equivalents that might explain their thousand-year medicinal use.

Of course seahorses contain various molecules and compounds that are already known to science and it is through these that some scientists are desperately searching for any signs of therapeutic benefit. A recent textbook of Chinese herbal medicines that includes seahorses—not all do—provides a list of their chemical constituents, mostly amino acids and fatty acids, including cholesterol, plus a few antioxidant pigments like astaxanthin and minute traces of metals, including iron and zinc. But none of these get to the bottom of how or why seahorses might work as medicines. The yellow seahorse is said to contain testosterone and the organic acid taurine, which is also found abundantly in bulls' testicles and is added, in a synthetic form and high concentrations, to energy drinks and muscle-building food supplements. Another TCM textbook refers vaguely to experiments in which female mice were fed an alcohol extract of Kellogg's seahorse, causing their ovaries and uterus to grow larger than usual. The unspecified dose of seahorse also apparently led to a prolonged estrus period in female mice and prompted an estrus period in castrated male mice.[31]

In 2003, a team of scientists from Sun Yat-sen University in Guangzhou in southern China conducted an in-depth study of seahorse genetics in the hope of discovering some clues as to their medicinal effects. They came up with lots of ideas, identifying many parts of the seahorse genome that code for important substances, including some they think might explain the ability of seahorses to enhance male virility.[32] Seahorses have been found to possess the gene for a hormone that influences the nervous and immune systems; some scientists also think this hormone is implicated in sperm production and fertility.[33]

Whether or not seahorses really do contain active chemicals in quantities large enough to have any effect, such studies may actually be missing the point. Chinese herbal medicines are based not on the direct effects of individual ingredients but on the bespoke cocktail of ingredients selected on a case-by-case basis. Isolating and testing specific compounds is, many argue, an unfair test. The problem is the severe lack of clinical trials that examine complex tailor-made drugs without dismantling the framework of TCM on which they rest. Researchers at the Peninsular Medical School in the UK searched through electronic databases and found over a thousand studies that claimed to test the effectiveness of individually prescribed traditional medicines. Out of all those studies, only three were properly randomized with double-blind tests, where both the doctor and the patient have no idea whether they are using the real treatment or a harmless placebo. One clinical trial investigated the use of Chinese medicines to alleviate the toxic side effects of chemotherapy. A traditional Chinese doctor consulted each patient and blindly administered the treatments, including a placebo, in the form of an herbal tea. The findings showed that the herbal treatments and placebos were equally ineffective in reducing the toxicity of the chemotherapy.[34] Far more studies like these are needed if a meaningful answer to the question "Do Chinese herbal medicines work?" is ever to be found.

A clinical basis for seahorse medicines may one day be found, or it may not. One thing we already know for certain is that some seahorses are on the ever-lengthening list of species that face an uncertain future. Six seahorse species have earned the dubious status of being labeled "vulnerable" to extinction by the World Conservation Union.[35, 36] For all six, the trade in traditional medicines is thought to play a large part in their threatened demise. One other species, the Cape seahorse, isn't involved in the medicine trade but, due to its miniscule population and deteriorating habitat, carries the more imperiled tag of "endangered." So little is known about all the other species, they are tagged with a noncommittal question mark or "Data Deficient." Setting aside for a moment any ecological and moral arguments for seahorse conservation, if TCM practitioners wish to see the long history of seahorse medicines continue, something must be done to ensure their persistence in the wild.

Since the time when seahorses were first used to restore yang, the traditional medicine trade has, as we might expect, radically changed. Instead of practitioners in China relying largely on locally available or even homegrown ingredients, now both supply and demand of herbal medicines have gone global. The international exchange of herbal remedies has grown into a multibillion-dollar industry. Even though western medicine is prevalent throughout China's urban populations, most people remain loyal, to some extent, to herbal remedies, using a combination of both to suit their needs. Centuries-old herbal retailer Tongrentang has established itself as one of the most valuable brands in China alongside Microsoft, Coca-Cola, and Nokia.[37] And the exotic aromas wafting through the doors of traditional pharmacies are no longer confined to the streets of China, with herbal outlets opening up in Chinatowns and malls in most major cities around the world, including New York, Paris, and London. As a consequence, the demand for seahorses is greater than ever.

Of mounting concern for the wild status of seahorses is the emerging market in prepackaged herbal remedies, a venture that veers a long way from the origins of Traditional Chinese Medicines. Instead of visiting a TCM doctor, a growing number of Chinese people are buying self-prescribed remedies off the shelf or over the Internet. This has created a burgeoning market for seahorses destined to be pulverized, mixed with other ingredients, and sold as mass-marketed tablets, pills, or bottles of health-giving tonic.[38] "Seahorse genital tonic pills" also sold as "Hippocampus virility restorative pills" contain a long list of ingredients that includes seahorses, tiger bones, fur seal genitalia, and ground whole geckos. This mélange of endangered species is described as a "systemic and brain strengthener," used to treat a range of conditions including "low sperm count or weak sperm, impotence and spermatorrhea, debility of limbs, night sweat or day sweat."[39]

With roots buried deeply within Taoism and Confucianism, TCM considers that humans are closely connected to the influence of nature. Indeed, the human body is a microcosm of the natural world. But by creating a demand for rare and expensive species like seahorses, tigers, turtles, and rhinos, Chinese herbal medicines in the modern world risk upsetting the balance of nature that lies at the core of its own philosophy, a fear that some practitioners are beginning to realize.[40] Medicines made from endangered species are not the cheap, widely available drugs that make TCM vital for millions of people. They are a high-priced choice prescribed for various chronic but nonlethal conditions. If seahorse medicines were no longer available, no one would die. And those with money to spend on these expensive luxury medicines also have the luxury of choice.

One option that could help stem the growing demand for seahorses and other endangered species is for practitioners to seek out alternative traditional ingredients that have similar effects. For thousands of years, Chinese medicine has been evolving and adapting, with new ingredients adopted and others falling into disuse. Why, then, shouldn't modern practitioners and users of TCM recognize the need to avoid using species that are threatened with extinction? A survey of TCM practitioners based in the United States revealed a general lack of dependence on ingredients derived from endangered species and considerable willingness to use substitutes instead. Of 145 practitioners who responded to questions about seahorses, nearly two-thirds said they considered them to be of minimal importance as a medicinal ingredient and only two people claimed they were crucial to their practice.[41] Other ingredients that could potentially replace seahorses include horny goatweed, Chinese dodder seed, and English walnut seed and, for the more adventurous, human placenta.

Promoting alternatives to species that are currently endangered can, however, be a risky business that in the past has gone tragically wrong. Demand for African rhinoceros horn, including as a Chinese treatment for fevers, was so fierce that by the mid-1980s, the number of rhinos in the wild was plummeting. In a desperate attempt to save the few remaining populations, conservationists encouraged researchers to investigate whether the properties of rhino horn could be matched elsewhere in

Chinese pharmacopoeia. The answer to the rhino crisis seemed to be an exotic-looking species of Asian antelope, with delicate ringed horns and odd, bulbous snouts. Saiga horn was listed in the *Shen Nong Ben Cao Jing* as a salty, cold medicine that calms the liver and cools heat, a likely replacement for rhino horn. At the beginning of the 1990s, saiga were highly abundant. At least a million of them roamed the semiarid rangelands of Central Asia, enough to convince conservation groups that the saiga would tolerate hunting. It was decided that saiga horn would be actively promoted as an alternative to rhino horn and it wasn't long before a vast new trade erupted. What the conservationists hadn't anticipated was the politics of the region and the speed at which both legal and illegal markets would emerge. To make matters worse, hunters and poachers quickly discovered how easy it was to round up and shoot the male saiga, the only members of the species with horns. This left behind a shrinking population with a catastrophic female bias; soon there were not enough males to fertilize all the females, and birth rates crashed. Since their introduction to twentieth-century medicine markets, the number of saiga has nose-dived by ninety-seven percent: Of the one million saiga alive in the 1990s, there are probably only around thirty thousand left. Nearly all of them are females.[42]

Another possible solution for suppressing the global desire for endangered species may lie outside the realm of traditional medicines. Amid the frenzy of international attention at the launch of the male performance drug Viagra, there was some hope among conservationists that it might eliminate demand for the unlucky endangered species that are believed to inflate sexual potency. One study in support of this theory pointed out that in 1998 sales plummeted in the velvet from Alaskan deer antlers and the genitalia of Canadian hooded and harp seals, both used as key ingredients in various libido-enhancing powders, tablets, and capsules. Trade in several other medicinal ingredients, including seahorses, also diminished around 1998 but had recovered again by 1999. Rather than being the direct result of the release of Viagra, a more likely explanation for the dynamics of the trade is the Asian economic crisis, which could have triggered the observed downturn in many imports into Asia, not just endangered species.[43]

It seems that after hundreds of years, it will take something more than a little blue pill to cure the world of its unsustainable addiction to dried seahorses. Even a 1998 report of worryingly high doses of lead found in prepackaged seahorse pills—which would surely dampen a man's virility—has not been enough to put people off.[44] The undying international demand for seahorses is the driving force behind a lucrative and thriving trade. At the latest count, at least seventy tons or twenty-five million seahorses were being plucked from the oceans every year, the majority destined to be roasted, crushed, and dissolved into traditional medicines.[45] And this may even be an underestimate. How is it, then, that so many rare camouflaged creatures are captured from their hidden homes on the seabed and conscripted into the global market place? To find out, we must turn to another species that has little to do with seahorses or herbal medicines.

Chapter Four

CATCHING POSEIDON'S STEED

*S*eahorses lead quiet lives, tucked away out of sight on the seafloor, sucking plankton and staying put, which means that unlike most other sea creatures, there has never been a special technique invented to grab them from their homes and bring them into our world. Seahorses don't huddle together in large scoopable shoals; they don't cruise the open oceans and become irresistibly drawn to blinking lights fixed to fishing lines; they don't care for the whiff of food laid inside an inescapable trap. Instead, nearly all the seahorses that end up being traded as traditional remedies are caught by accident by shrimp trawlers, the least selective means of fishing there

∞

Above: Dwarf Seahorse
Credit: Theodore Gill, 1905

is; all, that is, except for a few rare fishermen who attempt to earn a living from seahorses, plucking them one by one from the seafloor by hand.

Seahorses are the focus of a handful of small-scale fisheries scattered around the tropics. In the 1980s, children in the Moluccan Islands of Indonesia collected seahorses from seagrass beds, giving them to Chinese traders in exchange for sweets; the bigger the seahorses, the more sweets they got.[1] In the remote Galápagos Islands, Ecuadorian biologists have seen divers collecting seahorses and selling them to sea cucumber traders from Asia.[2] A few thousand seahorses are collected each year by scuba divers in Acapulco, on Mexico's Pacific coast, and small numbers are caught by diving operations in Costa Rica, Panama, and Peru. In the 1990s, there was a fishery on Tanzania's Mafia Island that targeted seahorses at night.[3]

The world's best-known dedicated seahorse fishery is at Handumon, a small village much like many others in the Philippines, where a cluster of wooden houses perch high on stilts on a hummock of dusty sand called Jandayan Island. A short way offshore from Jandayan runs an unusual double barrier reef, one of only six known in the world. It is from the rich mix of habitats that weave through the parallel bands of reef, known as Danajon Bank, that the residents of Handumon must make a living. Where land meets sea, mangrove trees dip their roots in shallow waters, creating playgrounds for crabs and nurseries for young fish; farther offshore, coral patches mingling with clumps of seagrass and sargassum weed create homes for seashells, squid, octopus, and seahorses.

In the 1960s, a shellfish trader arrived at Handumon asking if there were any seahorses for sale. At first, children plucked them from seagrass beds close to the village and, before long, adults joined in. Fishermen who caught fish from the reef using spears and nets began fashioning goggles from the bottoms of glass bottles and diving fins from planks of plywood so they could dive for seahorses. They ventured out at night, apparently spying the seahorses' eyes, like a cat caught in headlights, shining brightly in the dancing light of a kerosene lantern hung from a wooden canoe on the surface. They became known as the lantern fishermen.

Thirty years later, seahorses had become an important part of the Handumon economy. Several traders had set up shop on the island and forty percent of residents were earning much of their income from seahorses. The idea had caught on, and lantern fishers were appearing all along Danajon Bank. By then, seahorse stocks were already beginning to show signs of wear and tear. Back when they first collected seahorses, fishers expected to find at least fifty or even a hundred specimens during a single night at sea. By the mid-1990s, a catch of twenty seahorses was deemed to be a good one and fishers would often return home empty-handed.

In 1993, the lantern fishers of the Philippines and their reports of disappearing seahorses caught the attention of an international team of biologists. They visited Jandayan Island and explored the nearby reefs, where they immediately realized that as well as being caught by local fishermen, the seahorses had an even greater problem to deal with: the breakdown of the marine habitats they call home. Danajon Bank's reefs were in a bad way and the seahorses, as well as the seahorse fishers, had little hope for the future unless the condition of their habitats could be improved. The scientists decided to set up an initiative to try and get the lantern fishers actively involved in protecting the natural resources they relied on. It was the beginnings of the world's first dedicated seahorse conservation program, Project Seahorse.[4]

Having gained approval and support from the village leaders, the research team held informal discussions with Handumon's fishermen, keen to hear their perspectives on the problems they and the seahorses faced. Everyone they spoke to blamed the poor health of marine habitats, and plummeting seahorse catches, on the destructive fishing of others. A crude way of catching fish is to fashion a bomb from plant fertilizer and toss it in the water. A lethal compression wave rips through the water, killing fish within a few meters' radius by rupturing their internal air bladders. Floating dead fish are picked from the sea surface, while down below a crater of pulverized coral is the enduring memory of a single fish bomb. The fish are usually small and their flesh macerated by the blast, making them fit only to be used as cheap bait. Another devastating fishing technique uses a squirt of cyanide solution from a plastic squeeze bottle held by a diver to temporarily stun, but hopefully not kill, large vibrant reef fish that have become favored delicacies in high-class seafood restaurants across Asia. Diners will pay a premium to individually select their dinner from a throng of live fish swimming in aquariums at the tableside. The problem is that back on the reef, the cyanide leaves behind a lingering, noxious miasma that extinguishes all other life. As well as proving unpopular with the artisanal seahorse fishermen, fish bombs and poisons are illegal in the Philippines and across Southeast Asia, and yet they remain popular tools.[5]

With a little support, the fishers of Handumon were eager to get involved with developing strategies to combat destructive fishing and stem the seahorse decline. They voluntarily filled out daily catch logbooks, offered their seahorses to be weighed and measured before selling them, and helped out with surveys to monitor the health of the local reefs and seagrass beds.

Following further brainstorming sessions, the fishermen

stepped forward with their own ideas of what to do. In August 1995, with the help of the village council, a thirty-three-hectare marine sanctuary was set up on the shores of Jandayan Island where all types of fishing were declared strictly off-limits. The idea, though not a new one, was to give parts of the seas a chance to recover, to provide a safe haven for species to grow and breed and hopefully, eventually, for their offspring to wander outside the reserve boundaries and help replenish other areas of battered reef.[6] Catches of small seahorses were donated to the sanctuary in the hope that a healthy population could be established. Villagers, including those not involved in the seahorse fishery, began to take great pride in the recovering status of their underwater sanctuary, taking turns to patrol the reserve boundaries to make sure outsiders didn't flout the new rules.

Experiments were also conducted with seahorse paternity wards. For a while, whenever fishermen caught pregnant males, instead of selling them to traders straight away, they were encouraged to put them in mesh enclosures fixed on the seabed close to shore. As soon as the babies were delivered and had drifted off through the cage bars to join their wild relatives, their dads could then be killed, dried, and sold. Although in theory this should have helped to sustain seahorse populations, in practice it proved too difficult to maintain the cages and persuade fishermen to hold off selling their catches until the males gave birth.

News from Handumon quickly spread. Soon, marine sanctuaries began emerging across the region with the help of Project Seahorse and a new regional alliance of fishers called KAMADA (The Alliance of Lantern Fishers on Danajon Bank). Spawned from the flourishing Handumon project, in 2002 KAMADA began working with local government, taking steps to stamp out illegal cyanide and dynamite fishing on Danajon

Bank and encourage more communities to set up their own marine management initiatives. After ten years, Project Seahorse and KAMADA had helped set up twenty-two marine reserves. Each one is relatively small, between ten and fifty hectares, but together they add up. Some villages have elected to set aside a substantial proportion of their nearby reefs. It is still too early to judge the value of most of these sanctuaries since the benefits they are creating for local habitats and livelihoods have yet to become fully apparent. But as long as their guidelines are well enforced—something that Project Seahorse and KAMADA are working on by providing guardhouses and patrol boats—the marine reserves ought to thrive. Plenty of studies from around the tropics have compared what lives inside versus outside of marine reserves, and so far they have demonstrated without doubt that protecting segments of the seas leads to major improvements for both wildlife and fisheries.

Even with these mounting efforts for protection, recent surveys indicate that the reefs of Danajon Bank are still in a dire state. Instead of a high cover of live coral—branching trees and shrubs, foliose plates like giant dining tables and massive boulders that look like brains—the reefs are littered with small nubbins of broken dead coral, the unmistakable fingerprint of bomb fishing. With all these loose fragments rolling around, new corals have no chance of establishing themselves; wherever coral larvae try to settle down and stick to the reef, dead rubble acts like sandpaper, scouring them off. Fish abundance is also alarmingly low, although a small glimmer of improvement is beginning to show. Twice a year for six years, biologists have swum through several of Danajon's marine reserves, clipboards in hand, counting fish. By comparing these counts to surveys in unprotected waters, it is obvious that certain groups of fish have been the first to seek refuge inside the submerged sanctuaries. After a nine-year fishing ban at Handumon, the fish at the top of the food web, predatory fish like groupers and snappers, were about twice as abundant compared to populations outside the reserve. At two other marine reserves, there are flourishing populations of colorful butterflyfish, which are choosy reef inhabitants, restricted to areas of healthy habitat by their coral-nibbling diet.[7]

But recovery of seahorse populations is going to take time. Scientists have yet to observe any marked changes in the numbers living inside the Danajon Bank protected areas. But there is one good sign from Handumon. An intensive survey that began in 2001 shows that average seahorse size is rising. In 2005, most of those living in the reserve were around fifteen centimeters long, compared to sixteen centimeters one year later.[8] It may only be a small step, but at least it is in the right direction toward seahorses that live longer and produce more offspring.

From a human perspective, Project Seahorse's initiative in the Philippines has been a huge success. KAMADA is now the voice of more than nine hundred fishing families from across Danajon Bank. Project Seahorse is also nurturing the next generation of conservation-minded fishers by funding more than thirty local children through high school in return for their services as "environmental ambassadors." The Handumon model is likely to work well in other locations where small communities are involved in collecting seahorses by hand, but only if key aspects of the original project are emulated. From the very beginning, Project Seahorse has been forward-thinking in embracing the concept of "community-based conservation," buzzwords that are popular now that protecting biodiversity is increasingly coupled with protecting local people's livelihoods. It is a move away from the "fine and fences" approaches of the past when indigenous people were deemed incompatible with valued wildlife and were sometimes forcibly relocated away from parks and protected areas. The Handumon community was consulted from the word go, and Project Seahorse works tenaciously to ensure that fishers gain a sense of ownership and stewardship over their marine resources and do not become reliant on the external crutch of foreign funding and support. Unfortunately, though, such careful and conscientious artisanal fishermen are the exception rather than the rule when it comes to supplying seahorses for the international medicine trade.

The bad news for seahorses is that they commonly live in the same parts of the seabed as shrimp—a highly desired commodity caught to furnish our seafood platters. A large proportion of the estimated twenty-five million seahorses traded each year are caught by fleets of trawlers when they go in search of the seahorses' crustacean neighbors. Large nets of ultrafine mesh are dragged across the seafloor, gathering up anything that happens to lie in their way. Not much escapes the attention of a passing bottom trawler. They are such an effective way of combing the depths that trawlers were used in 1996 to retrieve debris from TWA flight 800, which crashed into the Atlantic Ocean shortly after taking off from John F. Kennedy airport.[9] When they are deployed to catch their intended target, the nets of shrimp trawlers bring up a lot of other sea creatures that fishermen do not want, so-called bycatch. Shrimp trawlers snag more bycatch than any other type of fishery in the world. Accompanying every kilo of shrimp is an additional five to ten kilos of unwanted sea life: Mountains of corals, sponges, starfish, seashells, and young, unpalatable, unsellable fish are all thrown back into the sea—dead or dying.

Seahorses are routinely pulled from the piles of incidental trawler trash. In a single day, a trawler may catch just a few seahorses, sometimes none, sometimes a handful. But with hundreds of boats in hundreds of fishing ports in dozens of countries, this quickly adds up to millions of seahorses caught each year.[10] In Hernando Beach, Florida, a small fleet of thirty-one trawlers ventures out at night to catch pink shrimp from shallow seagrass beds. Each year they also catch around seventy-two thousand lined seahorses from a population of unknown size.[11] Inside the nets and on the decks of shrimp trawlers all around the world the same story is played out a thousand times over, day after day, night after night.

One such shrimp trawl fishery operates from Cua Be, a small fishing community clasped within a knot of narrow cobbled streets that flank the seaside city of Nha Trang, in southeast Vietnam.[12] At around four o'clock each afternoon, as the sun begins to slip behind layers of smoky mountains, a fleet of ten-meter-long wooden boats fills the air with the deep *bub-bub* of diesel engines. A hundred and seventy red flags with yellow stars, one per boat, flap furiously in the onshore breeze as the convoy departs for a night on the South China Sea. After an hour of motoring toward a denim blue horizon, engines idle and bottle green nets are hauled overboard, followed by paired metal otter boards that shear through the water like airplane wings, holding open the net mouth. Gears are once again engaged and the trawlers begin a slow haul, plodding forward at walking pace, tugging their nets across the seafloor twenty meters below. By the time the skies darken, the fleet can be seen from shore as a constellation of distant bright pinpricks against the seamless black of sea and sky.

The fishermen must wait for four long hours before they find out what they have caught from the unseen depths. Eventually a rusting winch pulls the net back up and a bolus of sea life is unleashed on the deck in a deluge of wriggling slime. Much of what emerges are crustaceans of one kind or another, some of them shrimp, a lot of them small crabs, none larger than a fist and the majority already dead, limp legs splayed. A few survivors fight their way to the top of the pile, claws wide and threatening as they wrestle and snap at each other. While nets are thrown in for a second haul, fishermen set about untangling the pile of oceanic debris: one basket for crabs, one for shrimp; one for silvery thumb-size fish; one for the corpses of miniscule squid and the occasional octopus still alive and throbbing like a human heart. Everything else is scraped up with cardboard paddles to be sold as "trash fish," the deep-sea wreckage that is not fit for human consumption but is fed instead to lobsters. Lobsters are farmed along the Vietnamese coast to meet demand from the wealthy elite and foreign visitors. Some fishermen specialize in catching tiny juvenile lobsters in fine nets at the shoreline and selling them to lobster farmers for ten U.S. dollars apiece. This can quickly add up to a small fortune, considering the average monthly wage in Vietnam is around sixty dollars.[13] The baby lobsters are then fattened up on the dregs of the sea. Up and down the coast, trawlermen fear that the recent arrival of a lethal lobster disease to Vietnamese waters could bring an end to the seven thousand Vietnamese dong they earn for every kilo of trash fish scooped from their nets, a grand total of around forty U.S. cents.

By four in the morning, the Cua Be fleet is returning to port, where mounds of crushed ice are unloaded onto the dimly lit streets, and fish traders await the night's catch. Among piles of trash fish, shrimp, and crabs, the fishermen will have found a few seahorses. There aren't many, only around twenty or thirty per boat per month, enough to earn a little extra cash every few days. Some skippers claim it is worth risking their nets on jagged rocks surrounding offshore islands where many seahorses lurk. Clusters of women sitting on pavements mending piles of nets with long needles hint that this might be true. Depending on the size and species, seahorses are sold to local buyers for around twenty or thirty thousand Vietnamese dongs apiece—a maximum of two dollars per seahorse. The only people to make decent money from seahorses in Vietnam are the buyers and traders. With their tails stretched out straight, rows of dead seahorses are laid out to bake in the sun for two days, like sunbathing vacationers. Traders will often bleach the skins of dark seahorses and try to pass them off as the paler specimens that ancient Chinese texts claim make higher-quality medicines. For a Vietnamese trader, the prize catch is a Kellogg's seahorse, larger than a human hand and carrying a retail price tag of more than thirty dollars a pair.

Once they have been caught and dried, seahorses follow erratic and sometimes unexpected trade routes around the globe. With such lightweight carcasses, they are inconspicuous to transport, easy to tuck away beside other products, often piggybacking on larger, more lucrative trades in dried sea cucumbers and shark fins. Vietnamese seahorses travel north through Hanoi and across the border to Guangxi Province in China.[14] Seahorses headed farther afield travel by air. In the 1990s, the large population of Tamils living in Singapore were offered free flights to visit family in India in return for filling up their suitcases on their way back with dried seahorses caught in the waters off Tamil Nadu.[15]

The seahorse trade spans at least seventy-seven countries. Most of the seahorses that enter the international marketplace were born on coral reefs and seagrass meadows in Indonesia, the Philippines, Thailand, and Vietnam. Many also emanate from India and Mexico. Nations with large Chinese populations have the most insatiable seahorse appetites and form the central hubs for the trade, sucking them in from around the planet. The latest entrants to the trade are seahorses from the Red Sea. Import records from 2006 show that Taiwan, a big seahorse consumer, received over half its annual consignment from Egypt.[16, 17]

Not all dried seahorses are destined for the medicine trade. Particularly in Latin America, seahorses are given a more whimsical use, sold as souvenirs and curios to tourists in seaside resorts.[18] With a ring poked through an eye socket, a dried seahorse makes an intricate but flimsy key chain. They are transformed—in an echo of ancient storytelling—into dragons with a glued-on pair of wings. Or they join an everlasting ocean scene alongside a handful of seashells and starfish entombed in the see-through blue resin of a toilet seat.

So what effect does trawling have on seahorses? There is little doubt that trawlers can inflict unchecked collateral damage on the seahorses' undersea homes. It has been estimated that every year at least half of the world's continental shelves, the parts of the sea shallower than 150 meters where all the seahorses live, are scoured by trawlers at least once—if not many times over. The equivalent of trawling on land would be for a hunter to hook up a bulldozer to a giant net five meters wide and twenty meters long, weighed down by a hundred kilograms of lead, and grimacing with a row of giant metal teeth. Setting off into an area of forest, the hunter drags his net along, trampling over and uprooting every tree and shrub in his path, scouring a trench fifteen centimeters deep into the soil as he goes—all because he wants to catch squirrels. And unless they can scamper away quickly enough, he certainly will catch squirrels along with all the other animals, lichens, flowers, and toadstools that live in the forest, most of which he will pick from his nets and throw away dead.

Although the seabed isn't covered in trees and shrubs, it is far from being a featureless flat plane, invulnerable to physical damage. Much of it is littered with boulders, cobbles and pebbles, living reefs and seaweed forests, and humps and hillocks of sediment created by burrowing animals; all of them provide crucial three-dimensional homes for other bottom-dwelling ocean life. A passing trawler does away with all that. Thousand-year-old deep-sea corals are brought gasping to the surface, lush green seagrass meadows are plowed up. The paths that trawlers tread across the seafloor can be seen from space, a crisscross of lines that take decades to fade. Much of the seahorses' habitat is far from a safe place to live.[19]

Looking at the seahorses' way of life, we can easily guess what happens when trawlers and seahorses collide. In terms of their biology, they are prone to all kinds of disturbances. They are too slow to outswim a passing trawler net, and if one member of a seahorse pair is caught, the widowed seahorse will be forced to take time out from the busy breeding schedule and hunt around for another mate.

Their reproductive strategy renders seahorses especially vulnerable to the impacts of shrimp trawlers. The number of offspring produced each year by a single seahorse varies between species: Larger males tend to have room for bigger broods, but on the whole, seahorse offspring are far from bounteous compared to the average fish's. Pacific and slender seahorses can give birth to more than a thousand young at a time, but a typical brood size for most other species is several hundred; tiny dwarf seahorses can only manage a meager five or ten offspring at once. Although each baby seahorse is well developed when it is born and has a better chance of survival than one of the three million eggs laid by a female Atlantic cod, their relatively restricted numbers nonetheless limit the ability of any survivors to replenish nearby areas that have been stripped bare of sea-horses by a trawler.[20]

The significant investment of time and energy into each baby by pregnant males and doting dancing females adapts them well to a calm sedate world where the daily risk of death is minimal, a way of life that ecologists refer to as K-selected. In contrast, r-selected species thrive in an unstable erratic environment, where life is less certain, chances of an early death are high, and all that really matters is to reproduce as quickly and as copiously as possible. K versus r is the essential difference between a hundred-ton whale and a five-gram minnow. Fast-growing r-selected species, including shrimp, often cope much better with being hunted, fished, or trawled by humans than do the K-selected species. Biologists are increasingly witnessing shifts in the types of animals living in the heavily trawled parts of the seas. Where once there were long-lived, slow-growing, slow-to-mature K species in these areas, now ephemeral r species are taking their place.[21] In parts of North West Australia where trawling is pervasive, high-value K species such as emperors, snappers, and groupers are being replaced by less commercially valuable r species, such as bream and lizardfish.[22]

To measure the actual impact of trawling on seahorses is an ongoing challenge. Surveying seahorse populations is as laborious as trying to hunt for inch-high toy soldiers that are dressed in green camouflage gear and have been thrown haphazardly into acres of long grass. Then to work out how much they diminish over time when they are fished is an even harder task. One way to tackle the problem is to focus on small-scale experiments. Studies in Portugal have shown that lightweight fishing nets, called beach seines, can have a significant impact on seahorses, although sometimes not in ways we might expect.

The Ria Formosa lagoon is a convoluted network of sand

flats, creeks, and channels that lie to the south of the Portuguese city Faro. The lagoon is home to two species of seahorse, the long-snouted and short-snouted seahorses, which take up residence in an estuarine neighborhood of seagrasses, seaweeds, sea urchins, and encrusting mosslike animals called bryozoans.[23] In 2001 and 2002, a team of seahorse researchers was granted rare permission to conduct experiments to investigate what happens when beach seines are dragged across small patches of seahorse-rich habitat. Since all other fishing is prohibited in the lagoon, the researchers had the luxury of complete control over which areas were fished and which were left alone. Twelve research plots were picked at random, each one twenty-five meters square, or around the size of six pool tables pushed together. Some were left entirely unfished, some were fished every month for two years, and some were fished for the first year and left to rest for the second.[24]

At the end of the two-year experiment, some clear patterns began to emerge. In the plots that were fished for two years, the number of long- and short-snouted seahorses was predictably low. The plots that were left untouched had, as expected, a steady and high number of both type of seahorse. It was the plots that were fished for the first year and left alone for the second where things started to get interesting; the two seahorse species responded differently. The long-snouted seahorses did much better when fishing stopped, the numbers tripling from 2001 to 2002, while at the same time, the short-snouted seahorses did the opposite. Following the cessation of fishing, the number of short-snouts dropped by almost half. It would appear to be illogical for two seahorses to react so differently when they appear to be so similar; they grow to around the same size, the height of a human hand span, they reach maturity at around the same age, and they produce roughly the same

number of offspring each year. It is only by observing them in
the wild that a key difference becomes apparent. Long-snouts
tend to stick to areas of lush, dense vegetation with lots of sea
urchins and bryozoans to hide among; meanwhile short-snouts
are braver, choosing to live in less crowded, more open sandy
areas. The two species' contrasting reactions to fishing might
have more to do with their distinct habitat preferences than with
the direct impact of being caught. When fishing stops, vegetation and mobile animals can quickly recover, making conditions favorable for the long-snouts, which like to hide away, but
this isn't so good for the bold short-snouts. All in all, this means
that there is unlikely to be a one-size-fits-all remedy for keeping
the various seahorse species happy in human-dominated seas
and estuaries. Some like to be left alone, while others may benefit from having vegetation cleared away occasionally to give
them a bit more room.

While individual studies like this give us an idea of how different seahorses deal with fishing pressure, they form only a
small part of a vast global picture. There is one solution to
the problem of assessing the seahorses' worldwide status that
doesn't even involve getting wet: Talk instead to the people who
catch seahorses and make a living from them; ask if they have
noticed any changes. Scientists have interviewed fishermen and
traders around the world and have revealed a recurring message
of the seahorses' downfall. In recent years, time and again,
trawlermen have reported that they find fewer, smaller seahorses
snagged in their nets than they used to. Indian fishermen say
that, in just three years, their catches crashed by three quarters.
In Mexico, catch rates are down to only five percent of what
they were twenty years ago. Between 1985 and 1995, fishermen
in Indonesia claimed that seahorse populations declined by
seventy percent. Vietnamese trawlermen reported a five-year

drop in seahorse catches of between thirty and sixty percent. And in 1996, seahorse buyers in Da Nang, central Vietnam, were forced to shut up shop when local supplies of seahorses dried up.[25]

Despite numerous reports of localized declines, the seahorse trade marches on and increases by around ten percent every year. This relentless pace of expansion is sustained by the tentacles of the trade reaching farther and farther around the world, encompassing more countries and more trawlers. Every year, more boats join the global trawling fleet, many subsidized by governments keen to keep foundering fishing industries alive. More seahorses are caught and more are kept, where once they were thrown back. Even so, from Hong Kong and Taipei to London and Los Angeles, consumers are so hungry for seahorses that demand still far outstrips supply.

This unceasing global demand for seahorses is driven, in part, by the human inclination to want what we can't have. The "water-diamonds" concept was originally based on the apparent contradiction between the low value of water, which is abundant but vital for human survival, and the very high value of the large diamonds we make into jewelry, which are rare and serve no practical purpose (although smaller diamonds are, of course, used in drill bits). Essentially this means that when commodities become scarce, demand goes up. This is especially true for luxury items, since rarity enhances that precious feeling of exclusivity, and as supplies dwindle, some people will be prepared to pay more and more for the desired product—a feedback loop that can mean the higher the price, the more appealing a product becomes. Wild species are known to suffer the same phenomenon. In the highlands of Papua New Guinea, butterfly collectors will pay more to local villagers for finding the rarest specimens, irrespective of their size or beauty. In Japan, medicinal flowers of Thunberg's geranium are found in two different colors, red and white, which vary in relative abundance across the country. In the west, the red flowers are common and seen as inferior in potency for treating stomach upsets compared to the rarer white flowers. The opposite is true in the east, where people favor the rare red flowers over the common white varieties.[26] The same capricious rules of human desire, mingled with the economic forces of supply and demand, also apply to seahorses.

If global assessments are right and the annual harvest of seahorses is already too high for wild populations to sustain, then what can and what should be done about it? Some argue that active management is vital since, unlike other fisheries, there are no inherent checks and balances that kick in when seahorse numbers begin to fall. Though such a thing is rare, a fishery that focuses on a single species will stop operating if—or more probably when—their intended target becomes too rare and difficult to find. When catch rates drop, there comes a point when the price of fuel and wages for crew outweighs the money made from the few fish that are caught and so it makes no financial sense to continue fishing. The fleet is forced to hunt for other fish, move elsewhere, or close, and the original targeted species is considered to be economically extinct even though a few may be left in the wild. This has happened repeatedly in fisheries targeting orange roughy, a member of the slimehead family that inhabits the coldest, darkest depths of the oceans, where they live dawdling long lives. When ordering orange roughy, most seafood lovers have no idea the fish on their plate may have lived for as long as a hundred and fifty years before it was caught. These fish mate for the first time at the age of twenty, gathering at giant, steep-sided submarine mountains called seamounts. Fishing boats flock to these same seamounts, where they scoop

Drawing of a Rainbow
Serpent rock painting
discovered on the
Arnhem Land plateau
in Australia in 1993

Credit: Jane Smither, Paul S. C. Tacon,
and the Aboriginal Community of
Kakadu National Park, Australia

Ribboned pipefish

Credit: Todd Stailey, 2008,
Tennessee Aquarium

Neptune by Leonardo
da Vinci, c. 1504-05

Credit: The Royal Collection © 2008
Her Majesty Queen Elizabeth II

¶ In eodem libro vt supra. Monoceron est
monstrū marinū: habens in fronte cornu ma
ximū: quo naues obuias penetrare possit: ac
destruere ⁊ hominum multitudinem pdere.
¶ Sed in hoc pietas creatoris hūano gene
ri puidit: quia cum tardum animal creatum
sit: naues eo viso possunt effugere.

Seahorse and fishermen
woodcut from Hortus
Sanitatis, 1497

*Credit: Unknown, Strasbourg edition
printed by Johann Prüss*

Counterfeit
hippocampus brooch
from the Lydian Hoard

Credit: AFP/Getty Images

Pictish carved seahorses
from Aberlemno, Scotland

Credit: Helen Scales, 2008

Knobby seahorse

Credit: David Harasti, 2008,
www.davidharasti.com

Leafy seadragon *Credit: Todd Stailey, 2008, Tennessee Aquarium*

Ornate
ghost pipefish

*Credit: David Harasti,
2008, www.davidharasti.com*

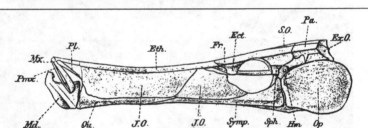

FIG. 1.—BROAD-NOSED PIPE-FISH (SIPHONOSTOMA TYPHLE). SKULL FROM SIDE SHOWING ELONGATION
OF FACIAL BONES, SMALL MOUTH AND JAW BONES, AND ABSENCE OF PREOPERCLE. (AFTER SCHÄFF.)
B. O., BASIOCCIPITAL; *Ect.*, ECTETHMOID; *Eth.*, ETHMOID; *Ex. O.*, EXOCCIPITAL; *Fr.*, FRONTAL; *Hm.*,
HYOMANDIBULAR; *J. O.*, INFRAORBITAL; *Md.*, MANDIBULAR; *Mx.*, MAXILLARY; *Op.*, OPERCULUM;
Pa., PARIETAL; *Pl.*, PALATINE; *Pmx.*, PREMAXILLARY; *Pt.*, PTEROTIC; *Qu.*, QUADRATE; *S. O.*, SUPRA-
OCCIPITAL; *Sph.*, SPHENOTIC; *Sym.*, SYMPLECTIC.

A scientific drawing of a broad-nosed pipefish skull   *Credit: Theodore Gill, 1905*

Big-belly seahorse   *Credit: David Harasti, 2008, www.davidharasti.com*

Tiger tail seahorse holding on to a pipefish   *Credit: Helen Scales, 2008*

Banded pipefish

Credit: David Harasti, 2008,
www.davidharasti.com

Pairs of dried seahorses on
sale in Nha Trang, Vietnam

Credit: Helen Scales, 2008

Energy-giving sea-
horse tonic on sale in
Nha Trang, Vietnam

Credit: Helen Scales, 2008

Vietnamese fisherman
sorting through the
night's catch

Credit: Helen Scales, 2008

A Vietnamese trawler departing for a night on the South China Sea *Credit: Helen Scales, 2008*

One day's seahorse catch at Cua Be, Vietnam *Credit: Helen Scales, 2008*

A dead Kellogg's seahorse

Credit: Helen Scales, 2008

A lantern fisherman from
Handumon Island, Philippines

Credit: Robb Kendrick

An etching of the four original
seahorses that were put on show
at the London Fish House

Credit: Illustrated London News, *July 23, 1859*

A one-inch tall dwarf seahorse

Credit: Todd Stailey, 2008, Tennessee Aquarium

Weedy seadragon

*Credit: Todd Stailey, 2008,
Tennessee Aquarium*

Children at the Tennessee
Aquarium imagining they are
sitting on the sea floor

Credit: Todd Stailey, 2008, Tennessee Aquarium

Young tiger tail seahorses at a ranch in Vietnam clinging to an air pipe, and each other

Credit: Helen Scales, 2008

Researcher David Harasti diving in Manly Cove, Australia, with a White's seahorse

Credit: Jayne Jenkins, 2008

Twelve identical seahorse heads waiting to be installed on the Civic Centre tower, Newcastle upon Tyne, 1968   *Credit: Newcastle Libraries and Information Service*

Robert Crowther's seahorses from the "Animals" exhibit at the Zoology Museum, University of Cambridge, in 2008. The exhibit featured special-needs students' work that was inspired by the museum displays.   *Credit: Robert Crowther, 2008*

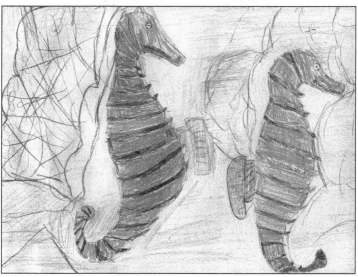

Bargibant's seahorse

Credit: Annelise Hagan, 2008,
www.annelisehagan.com

Denise's pygmy seahorse

Credit: Denise Nielsen Tackett,
All Rights Reserved

A juvenile of the Bargibant's seahorse

Credit: Denise Nielsen Tackett, All Rights Reserved

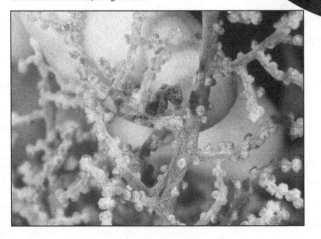

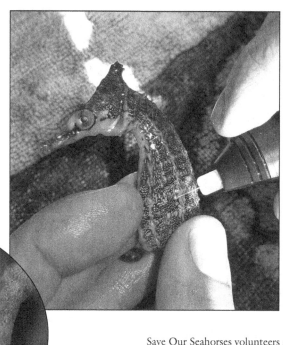

White's seahorse being tagged in Port Stephens, Australia

Credit: David Harasti, 2008, www.davidharasti.com

Save Our Seahorses volunteers surveying seagrass beds at the Pulai River Estuary, Malaysia   *Credit: Serina Rahman, 2008*

Thorny seahorse

Credit: David Harasti, 2008, www.davidharasti.com

Severns' seahorse, a new pygmy seahorse that was named in 2008

Credit: David Harasti, 2008, www.davidharasti.com

up thousands of tons of spawning fish in one go.[27] As soon as the colossal catch rates begin to slump, the fishing fleet moves on to the next seamount, leaving behind a few fish that will begin rebuilding the population, even if it takes them centuries to do so.

In contrast, trawl fisheries tend to take a mix of different species and there is no built-in mechanism that will divert the fishermen's attention and save the last few individuals from being caught. Even worse for seahorses is that they are deemed merely an added bonus and not the trawlermen's mainstay, so when their numbers inevitably decline, economic disaster doesn't strike and trawling will carry on regardless. There are still other fish to catch and still money to be made. In theory, trawlers could keep on catching seahorses until the very last one is gone. This is exactly what thrust another species, the barndoor skate, a flattened, shovel-nosed relative of sharks, to the cusp of extinction in the Northwest Atlantic. When they hatch, barndoor skates are around twenty centimeters long, almost too big to fit through a mailbox slot. In one sense, this makes the young skates tough survivors but it also means they are inescapably vulnerable to the trawlers that plough through their home range; the holes in the trawl nets range from seven to fourteen centimeters, so there is just no getting away for the baby barndoor skates. In the 1950s, one in ten trawls on the St. Pierre Bank off southern Newfoundland caught barndoors. Subsequently, catch rates tumbled and since the 1970s not a single one has been pulled out of a trawl net.[28] Similar trends were seen throughout the skates' natural range, and not because there were any changes in the type of gear used or the areas fished—there are simply no skates left.[29]

Realization that seahorses could share the same fate as creatures like the barndoor skate has prompted international efforts to protect them. Project Seahorse is not only responsible for spearheading seahorse conservation in the Philippines but has also shone the spotlight of attention at the seahorses' global plight. No matter where or how a seahorse is caught, whether it is snagged in a Vietnamese shrimp net or plucked by hand from a Philippine seagrass bed, it can no longer be freely traded across international borders. To pass legally through customs, all seahorses must be accompanied by a fistful of paperwork declaring that their removal from the oceans left no lasting footprint on wild populations. That goes for dead seahorses, live seahorses, or anything that contains any part of a seahorse. The decision to give seahorses international protection was a landmark vote that took place in 2002 in Santiago, Chile, at the twelfth biannual meeting of CITES (pronounced "siy-tees"), the Convention on International Trade in Endangered Species. Established in 1975, CITES aims to minimize the ecological reverberations that can propagate when wild species are traded as commodities in our human world. It is a voluntary convention, fueled entirely by goodwill and international diplomacy, which currently safeguards around thirty thousand species by controlling trade at two different security levels. No trade at all is allowed for species listed in Appendix I; these are extremely rare creatures, often the much-studied mammals such as big cats and primates, that are deemed to be teetering on the verge of extinction and could easily be pushed over the edge by exploitation for international trade. For species that are facing less imminent oblivion but are still in hot water, there is Appendix II, which allows limited trade to continue under strict regulations and controls.[30]

In the late 1990s, the director of Project Seahorse and one of the world's foremost seahorse experts, Amanda Vincent, conducted the first global assessment of the international seahorse trade. She gathered information from fishers, traders, and gov-

ernments around the world and discovered what she believed
to be compelling evidence that the survival of seahorses in the
wild was by no means secure, chiefly due to the trade in traditional medicines.[31] Further studies revealed that by the start of
the twenty-first century, the seahorse trade was booming. It
was decided that international conservation efforts were desperately needed.

What followed was a long and ultimately successful campaign to have seahorses properly assessed and considered for
inclusion in CITES Appendix II. This was the first time a commercially important fish species had been proposed for listing
by the convention, which until that point had focused largely
on animals traded as pets—mostly birds, mammals, and reptiles—
or rare plants sought by collectors. Also added to CITES in
2002 were the two largest fish in the oceans, the basking sharks
and whale sharks, which are threatened by the demand for shark
fin soup. Several CITES member nations vehemently opposed
the listing of seahorses and sharks, claiming that ensuring the
persistence of fish stocks should be the job of national fisheries
organizations, not international conventions; these countries
stand in the way of ongoing attempts to list other threatened
marine species on CITES. Four nations—Norway, Japan, South
Korea, and Taiwan—are taking advantage of a loophole in the
convention and have opted officially to ignore the restrictions
on the seahorse trade.

Control of seahorse trade by CITES came into force in May
2004, and it is still perhaps too early to tell what effect the measures will have.[32] A major stumbling block for implementing the
new CITES regulations is deciding on how many seahorses
each country can trade while remaining within the bounds of
"sustainability."[33] It is challenging enough to count how many

seahorses live in a given area, let alone to predict how many seahorses can be taken away while still leaving behind a viable population. Most seahorse-catching nations either have limited resources or lack the political will—or both—to divert attention to the tedious assignment of monitoring and properly managing the seahorse trade.

CITES has suggested a temporary solution based not on the number of seahorses that can be traded but on the size. With the help of seahorse biologists, a magic seahorse size of ten centimeters was identified, as measured from the top of spiny crown to the tip of curling tail. It may sound like a neat round number plucked from the air for convenience, but ten centimeters is in fact a crucial size at which virtually all seahorses, except for the very largest and smallest species, have reached sexual maturity. Ten centimeters is the smallest size at which most males have been found with babies in their pouch. By leaving behind specimens that are smaller than this, the trade should only take seahorses that have already had a chance to reproduce and cast their offspring into the next generation.[34] Most important, for now, this guideline is relatively easy to enforce. Fishers, traders and customs officers need only know how to measure seahorses accurately in order to determine which ones can, and which ones can't, be traded.

Even though CITES regulations apply to the whole seahorse genus, it is still a requirement of the convention that individual species being traded are identified, a major undertaking for customs officers who don't normally have a trained taxonomist watching over their shoulder. However, help is at hand in the form of the genetic tools that were originally developed to help solve academic disputes over species identity. Researchers in California demonstrated that it is possible to extract DNA from small clippings taken from any dried seahorse, except for those that have been bleached or crushed and made into pills. In 2005, they took samples from seahorses in curio stores and Chinese pharmacies in the San Francisco Bay Area and used the DNA to identify which species were present. It was quite a surprise to discover that most of the seahorses were not—as was expected—from Asia but were actually from much closer to home. Previously, Pacific seahorses were only rarely traded, but now that has changed; two out of three seahorses on sale in California in 2005 were Pacifics.[35] Regular genetic screening would not only provide information about which species are in the trade, but it could also pinpoint on a map where the seahorses come from. The troubles that taxonomists face when a single species is genetically diverse can now be put to good use. With a simple genetic test, it is now possible to work out whether a three-spot seahorse came from the Philippines or Papua New Guinea, and a lined seahorse can be matched to a certain region of the Atlantic coast of North or South America. Knowing which particular corners of the seahorses' wide kingdom are being plundered most heavily is crucial if plans for a sustainable trade are to work.

There are two other problems that CITES has stirred up in the international seahorse trade. Some countries take CITES regulations more seriously than others and have national laws banning trade in any species listed anywhere in the convention. Such legislation brings the legal seahorse trade to a grinding halt, including in the Philippines, where thousands of fishers across the country have officially lost an important source of income. Nonetheless, the trade hasn't stopped—it has merely been driven underground.[36] Project Seahorse discovered that since 2004, new buyers have emerged across Danajon Bank, and seahorse prices are increasing. The trade is alive and well and seems to be responding to the incessant growth in global demand. Filipino fishermen are at least sticking to the minimum size limit. More than ninety percent of the seahorses that are still being caught are larger than ten centimeters, suggesting that KAMADA's campaign to promote awareness among fishing communities continues, to some extent, to be a success. Project Seahorse and KAMADA are putting pressure on the Philippine government to amend the Fisheries Code so their work toward low-impact seahorse fisheries can keep going. But until the seahorse trade is once again made legal—and ways are found to make it truly sustainable over the long term—there may be little hope for the seahorses and the seahorse fishermen of the Philippines.

Another issue that CITES has limited powers to deal with is the unintended decimation of seahorse populations. No matter what goes on at international borders, trawlers will still trawl, and seahorse capture will continue. Even if fishermen don't bother keeping and selling their bycatch, there are only slim chances of survival for a trawled seahorse if it is returned to the sea. The majority are already dead by the time they are hauled up after a long journey in a trawler net. In shallower fisheries that soak their nets for less than an hour, many seahorses do make it on deck alive. In Florida's Hernando Beach trawl fishery, less than one percent of snagged seahorses die during the towing and sorting process and most are thrown back overboard.[37] But many get badly damaged, losing part of their delicate tails (which they cannot grow back) and with it their dexterous tool, especially important for males that need their tails to wrestle with competitors over mates and to hold on tight to seagrass blades when their bellies are laden with babies. The proportion of seahorses that can tolerate such impairment remains unmeasured, but it is unlikely to be high; of four dwarf seahorses caught in short, five-minute trawls in Tampa Bay,

Florida, only one was still alive in an aquarium tank thirty-six hours later.[38] If seahorses can't survive being caught incidentally in trawl nets, is their capture avoidable in the first place?

The damage caused to marine wildlife and habitats by trawling fleets no longer goes unnoticed, but it continues to leave conservationists and fisheries scientists scratching their heads for a solution. Escape hatches can be fitted to shrimp nets to curtail the death toll of large charismatic air-breathers. Snagged turtles and dolphins push open these aquatic cat flaps and swim to freedom, leaving behind the rest of the catch.[39] Other gadgets are also being introduced to reduce the bycatch of unwanted species like sea snakes, sharks, and other fish that have no market value or allocated quota. These are low-tech devices, made from squares of widely spaced mesh, that make use of the different ways fish and shrimp react to being caught in a trawl net. Fish usually herd together and swim along, holding their position within the net as it gets pulled through the water until eventually they grow tired and fall back. As the shoal becomes crowded toward the end of the net, panic sets in and the fish instinctively swim upward or sideways, escaping through the wide mesh panels. Shrimp, on the other hand, cannot swim fast enough to keep up with the net and instead tumble along to the bottom, missing the panels altogether. The size of the mesh panel determines the size of fish that can escape. But of course these devices won't work for seahorses since in many ways they are similar to shrimp: They, too, are small, sluggish swimmers.

Perhaps the only way seahorses are going to avoid the nets of shrimp trawlers will be to halt trawling or at least for trawling to be banned from substantial parts of the seabed. But until that happens—if it ever does—the potentially dismal future that seahorses face can at least be diminished by a careful combination of the "top-down" initiatives like CITES and "bottom-up" approaches like the Handumon project. Project Seahorse has demonstrated for the first time that a tiny, almost invisible creature can be used as a motivating force to generate enthusiasm for protecting areas of habitat and all the other species that live there, too.

Project Seahorse strongly believes that small-scale seahorse fisheries can be sustainable and offer a long-term source of income to fishermen, even if we are still not sure exactly how to achieve such an outcome. What is certain is that lantern fishermen selectively catching seahorses cause far less damage than other fishermen using fish bombs, cyanide solution, and trawl nets. But well-managed artisanal fisheries will only ever supply a tiny fraction of the current global seahorse demand. And with loopholes in international conventions and ongoing incentives for poaching, a globally sustainable seahorse trade is a long way off. And there is no doubt that the trade in seahorses is here to stay. Demand is growing not only for dead seahorses to be made into medicines and trinkets, but there is also a price on the head of seahorses that are still alive.

Chapter Five

SEAHORSES IN GLASS HOUSES

The urge to scoop a fish out of the sea or a river or lake and have it live alongside us in a garden pool or in a fishbowl is not a new one. Outdoor fishponds have been around for at least two thousand years, with the inhabitants used as food, decoration, and sometimes assigned fortune-telling powers. The Romans were the first to bring fishes inside, keeping carplike freshwater fish called sea barbels in small stone basins so guests could be assured of a fresh meal. From such humble and practical beginnings grew the desire to create ornamental fish displays in our homes. The first mainstream underwater pets were freshwater goldfish kept in elaborate porcelain bowls in tenth-century China. The Japanese seized on the idea of breeding goldfish into a variety of flouncy

❧

Above: Shiho's Seahorse
Credit: Theodore Gill, 1905

shapes and striking colors. By the end of the seventeenth century, goldfish had arrived in Europe to great acclaim. Now, hundreds of years later, there is barely an aquatic creature that isn't held captive somewhere in the world, from giant whales to minute sea slugs, and among them are a growing number of seahorses.[1]

A craze for keeping marine aquariums at home first took hold in Victorian England, back in a time when studying the natural world was a fashionable and respectable pastime, a healthy outdoors pursuit that was perfect to discourage the wandering of idle minds and hands. People were swept up by a shared obsession with collecting, admiring, and cataloguing objects from nature; drawing room cabinets displayed the intricate whorls of seashells and oddly shaped fungi, seaweeds were dried and pressed into scrapbooks, birds eggs were neatly lined up in drawers.

For the Victorians, natural history offered a set of stepping-stones that neatly bridged the gulf between science and art. It embraced the need to count and identify while at the same time it was rooted in aesthetics and stimulated artwork, literature, and poetry, often emanating from a widespread belief that the beauty of nature was a testament of God's work. And there was much personal gain to be had. Thomas Henry Huxley, eminent Victorian biologist and grandfather of *Brave New World* author Aldous Huxley, wrote in 1854, "To a person uninstructed in natural history, his country or sea-side stroll is a walk through a gallery filled with fine art works, nine tenths of which have their faces turned to the wall. Teach him something of natural history, and you place in his hands a catalogue of those which are worth turning round."[2] Nature was free for all to enjoy and a cheap and easy way to fulfill a growing Victorian desire for gathering material possessions. It fed the imagination of the country, holding mass appeal to a broad cross section of society and gave rise to an army of amateur natural historians.[3]

Traipsing through the English countryside, armed with a collecting net and glass jars, may sound like a simple enough pursuit, but it changed the way a generation viewed nature and their place within it. Even while gentlemen explorers were returning from their romps across the empire, bringing with them stories of astonishing beasts and undreamed-of places, the general public's attention still did not drift overseas. There was an equally exotic and enthralling world much closer to home, and one that every man, woman, and child could discover for themselves. Anyone who wanted to could come eye to eye with an arena of limitless variety and beauty. And the more you looked, the more you would find.

The seashore quickly came to light as a place where nature's curiosities were especially bountiful. Up until the nineteenth century, the ocean had been an unknown, dangerous place guarded by dreadful sea monsters. But slowly, as the myths and legends of wild seas subsided, people began to relinquish their fears and instead were infused with an urge to seek knowledge from the depths. It was with a longing to understand the marine world that the Victorians were the first to try and tame it. They put glass walls around a slice of the ocean and brought it into their homes.

Much of the excitement that grew up around the British coastline revolved around the work of one man, Philip Henry Gosse. Born in Worcester in 1810, Gosse was the son of an impoverished painter of miniature portraits. At seventeen, with no university education or inherited wealth, he sailed to Newfoundland, took a job as a clerk, and in his spare time devoted himself to the world of insect collecting. For a few years he lived in mainland Canada, where he failed in his attempts to set up a commune and a museum of stuffed birds. On returning to England in 1839, Gosse brought back with him a finely tuned adoration for nature and began what would soon become a prolific career as a natural history writer. By 1852, after a bird-collecting trip to Jamaica and a trilogy of successful books, Gosse was suffering from what his son later described as "nervous dyspepsia induced by excessive brain work." Rustication was required. Gosse took his wife and young son away from the turmoil of London life to imbibe the clean tranquil airs of the West Country.[4] It was there that he soon found himself becoming infatuated by the sea and all the things that live in it. He spent his days in Devon pottering about the shore, peering under rocks, dabbling in tide pools, plucking creatures from the sea, and bringing them all back home in glass jars to scrutinize under microscope and hand lens. His subsequent book, *A Naturalist's Rambles on the Devonshire Coast*, painted a glorious picture of an unseen world that lay at the very edges of England. It was an unexpected world full of unimaginable creatures, a world of starfish and feather stars, squat lobsters and sea cucumbers, sea slugs that looked like a lemon cut in half, and scarlet sponges "as big as one's hand." There were watery gardens filled with daisy anemones and snake-locked anemones, smooth anemones and rosy anemones, all hiding amidst dense tufts of algae that grew like "thickets of prickly pears." Gosse wrote meticulous descriptions of all the creatures he captured in nets and chiseled from rocks. He measured and catalogued polyps and tentacles, fronds, spines, and bristles; he noted their diets and watched their behavior. "Stand still, you beauty!" he exclaimed to the prawn, "and don't shoot round and round the jar in that retrograde fashion, when I want to jot down your elegant lineaments!"

In 1854, Gosse was the first person to coin the term

aquarium, in his book of the same name. In *The Aquarium* he wrote further exhaustive descriptions of his coastal explorations illustrated lavishly in color and providing readers with instructions on how to create their own miniature ocean. It was a publishing sensation, propelling Gosse into the public eye as the country's foremost narrator of the coast and infecting the nation with his addiction to sea life.[5]

The word *aquarium* was a "neat, easily pronounced and easily remembered, significant and expressive term," Gosse proclaims in his book. He selected the word as the single neutral form of *aquarius* and an aquatic version of a vivarium, a tank that was already used to display land-based animals. Although Gosse is credited with inventing the word *aquarium*, there is some confusion over who was first to pinpoint a solution to the greatest obstacle for a smooth-running marine display: the need for endless supplies of fresh, oxygen-rich seawater.

Up until the 1850s, many people had taken sea creatures out of their natural environment and kept them alive in tanks for several days at a time. But none of them had ventured far beyond dashing distance of the shoreline and none had tried to establish a permanent setup. The concept of keeping a self-sustaining snapshot of nature in a sealed, man-made unit was pioneered by the surgeon Nathaniel Bagshaw Ward of Whitechapel in London. He was an avid fern collector but found that his exotic plants quickly became poisoned by Victorian London's filthy air. Then, in 1827, he made an accidental but profitable discovery. Several months previously, he had placed a moth's cocoon inside a glass jar with a handful of soil, hoping it would hatch. Eventually, having remembered his little experiment, he returned to find not an adult moth fluttering around but a couple of healthy ferns growing happily inside. A few spores must have been scattered in the soil, then germinated and prospered

in the sealed protective atmosphere. The Wardian case, a miniature fern hothouse made from a large upturned glass jar, became
an essential accessory in the homes of Victorian naturalists and
a vital tool for bringing ferns, orchids, and all manner of delicate plants back unscathed from distant lands. Dr. Ward tried
filling one of his cases with freshwater and aquatic plants, and
with the addition of a few fish he had taken a step closer but still
hadn't cracked the aquarium conundrum.[6]

The unsung hero of early aquarium studies is Anna Thynne,
wife of the Reverend Lord John Thynne, sub-dean of Westminster Abbey. She spent many years in the respectable surroundings of her London drawing room, studying the intricate forms
and intimate sex lives of stony corals.[7] Initially, she had supplies
of seawater brought from the coast, which she aerated by hand,
pouring it repeatedly between vessels to stave off stagnation. In
1849, she tentatively tried adding some seaweed to her tanks,
expecting them to help enrich the water with oxygen much like
the captive atmosphere inside Ward's fern cases. In doing so,
Thynne kept her corals alive for an unprecedented three years,
along with clusters of other marine animals and plants. It was
the first "balanced" marine aquarium in London and it attracted
many natural historians to visit Thynne's home, among them
perhaps Philip Henry Gosse himself and another man named
Robert Warington, a chemist at the Society of Apothecaries.[8]
Both men would be remembered far more enduringly than
Thynne for the parts they played in the world's first aquariums.

It was Warington who was the first to publish a paper outlin-
ing the theory behind a "balanced" aquarium. His paper "On
the adjustments of relations between the animal and vegetable
kingdoms, by which the vital functions of both are permanently
maintained" was presented to the Chemical Society in London
in 1850. Having experimented with a few Wardian cases, he

then took a twelve-gallon container and added to it two gold-fish, a handful of stones, springwater, a few pond snails, and a plant called *Vallisneria* or eelgrass. Putting all these ingredients together created a stable system that could look after itself without the need to aerate or exchange water. The fish survived on oxygen seeping from the submerged plants; in turn, the plants soaked up carbon dioxide, ammonia, and various other by-products of the fish's daily activities. The snails were added as a final touch to help "renovate" the tank, as Warington put it, munching their way through the green algae that grew on the glass tank walls. In many ways, these early aquarium studies paved the way for the scientific discipline of ecology, in which animals and plants are not considered as isolated units but as part of an interdependent system with linkages between living and nonliving things. Admittedly, this was a freshwater and not a saltwater aquarium, but soon, perhaps after visiting the Thynnes, Warington began experimenting with seaweeds, periwinkles, and sea anemones, buying supplies of English Channel seawater from oyster boats at Billingsgate fish market.[9] At around the same time, Gosse was also experimenting with marine tanks, possibly without realizing that Ward was carrying out his own investigations.

No matter who was the first to invent the marine aquarium, it was Gosse who was responsible for bringing them to the masses. Soon after *The Aquarium* was published, people began swarming to the shore, intent on procuring their own collections of marine wonders. Access to the British coast had recently opened up with the expansion of the nation's railways, and soon seaside holidays became all the rage. Aquariums offered amateur natural historians a whole new outlet for their desires to "ransack the world for visible and invisible wonders," as philosopher George Henry Lewes observed.[10] "England has become Gosse-ified, whole families at a time," *The Atlas* declared in 1856.[11] And by 1857, the satirical magazine *Punch* was poking fun at the hordes of seaside collectors. One cartoon depicted a beach littered with ladies hunting through rock pools, the limpetlike shapes of their upturned petticoats creating an additional pastime for gentlemen onlookers.

Trains returning from the coast in the mid-1800s must surely have been soaked in the stench of day-old seaside, with passengers clutching jam jars of sea life and slopping pails of water as they headed off to create their own oceans at home. For those who could not make it to the coast, city shops soon saw an opportunity to cash in on the latest trend. William Alford Lloyd established a roaring trade from his Aquarium Warehouse in Portland Road in London, where customers could buy seawater by the pint or gallon, choose from a range of mass-produced glass tanks with ornate cast-iron frames, and take their pick from fifteen thousand wriggling and swimming creatures gathered from the sea by teams of professional collectors.[12] Booksellers WH Smith took advantage of the smooth rail journeys, setting up stalls on railway platforms and offering a new generation of cheap literature to vacationers who read en route to their destinations. Natural history titles were best sellers, some combining instructions on how best to assemble natural history collections with advice on the more routine domestic matters of cooking, cleaning, and etiquette. Titles dealing with the oceans were especially popular. A cheaper version of Gosse's *The Aquarium* without color plates was released and other writers joined in with dozens of "how-to" aquarium guides and beach companions, all of them extolling the wild entertainment that could be pursued at the coast. "The wonders of the ocean do not reveal themselves to vulgar eyes," warned H. Noel Humphreys in his 1857 book *Ocean Gardens.* "None but the initiated can see the myriad miracles that each receding tide reveals on the ocean floor." Initiation, however, was not a mysterious affair with "dark rites to observe," but simply a few good books, including his own, that should be consulted before taking a trip to the seaside.[13] George Brettingham Sowerby professed in his 1857 version of *The Aquarium (A popular history of)* that it is "hardly necessary to recommend this new and popular method of studying nature, as a prolific source of amusement and instruction." But he nonetheless goes on to describe at great length the "picturesque miniatures of ocean life" to be found in tide pool grottos.[14] And in the elegant volume *Rustic Adornments for Homes of Taste*, Shirley Hibberd gave instructions on the indispensable kit for a day at the beach: "a geological hammer, a strong chisel tipped with steel, a neat hand-net, a large closely-woven covered basket, two stone jars, one large glass jar, or a confectioners show glass, and two or three small vials."[15]

Not content with drawing the public to the beach, Gosse took his aquariums to show the owners of London Zoo in Regent's Park. Immediately impressed, they set about building him The London Fish House, a steel and glass construction to hold dozens of tabletop and wall-mounted tanks. In May 1853, the world's first public aquarium opened with more than four thousand specimens that Gosse himself had assembled from the shores of Devon and transported by train to London.[16] The public was thrilled. "The secrets hitherto known only to fishes and mermaids are laid open to all who choose to know them," reported the *Daily News*. The London Fish House was "so crowded daily with its curious visitors," wrote H. Noel Humphreys, that it was "difficult to get a glimpse of the wonders of the 'ocean floor' and its zootypic denizens."[17]

The delights of seeing an ocean in the city caught on, with aquariums opening up across Europe and before long making their way to the other side of the Atlantic. The great American showman Phineas T. Barnum visited the aquarium at London Zoo during his 1855 tour of England as a bankrupt businessman giving motivational lectures on "The Art of Money Getting." Having sold the contents of his American Museum in New York to clear his debts, Barnum still worked for the new owners, seeking out novel curiosities to display. He returned to New York accompanied by two professional aquarium makers from London and a head full of ideas for a brand-new ocean exhibit, which was constructed alongside the dubious "FeeJee Mermaid" exhibit, an elaborate hoax assembled from the withered body of a monkey stitched to a dried fish's tail. America's first aquarium display had "no competition whatever in the western hemisphere" wrote Henry Butler, co-owner of the museum, in his 1858 book *The Family Aquarium*. It was, according to Butler, beyond dispute the "largest, most costly, most complete, and most elegant production of the kind on the face of the globe!"[18]

Butler's was the first marine handbook published in the United States and in it he urged the American public to follow in the footsteps of the English and embrace the joys of the ocean. "Turn to the aquaria!" he declared. "Ring up the curtain of your thoughts. There, indeed, is comedy and tragedy, broad farce and exciting melodrama." Now that purpose-built glass tanks were being manufactured by the museum, Butler pointed out, there was nothing stopping aquariums from becoming "the universal embellishment of the private parlor or the sitting-room, the conservatory or the garden." And it was not long before Americans did indeed catch aquarium mania. Specialist aquarium stores opened, new public aquariums sprang up at the Boston Aquarial Gardens and New York's Central

Park, books and magazines were published, and aquarium-keeping societies were founded.

Keen-eyed marine collectors in Europe and North America may have occasionally spotted a seahorse clinging to an eelgrass bed or bobbing past in a sandy bay. But perhaps because seahorses are naturally so rare, Philip Henry Gosse himself never reported finding one in any of his books. He briefly remarked on their chameleon-like eyes in *The Aquarium* under the subheading "Double Vision," quoting an observation made to him by a friend, Mr. Lukis, from the Channel Island Guernsey. Certainly, a more common syngnathid visitor to Victorian shores would have been the pipefish. Most Victorian marine handbooks and aquarium guides mention collecting and keeping the seahorses' slender relatives. Shirley Hibberd admitted in her *Rustic Adornments* that she was partial to the "lazy dreamy" pipefish because of the peculiar acrobatics they performed in her aquarium. "They assume every possible attitude except horizontal. And like the buffoons on the human stage, get laughed at for their pains." Gosse, however, was rather unsure in his opinion of these particular marine characters. At first he gave the impression of being rather fond of them. In *A Naturalist's Rambles* he gave an intimate description of a worm pipefish, which he kept in a tank for four weeks, writing, "the manners of this pretty little fish are amusing and engaging." He bestowed on it an air of intelligence as it spied him from between fronds of seaweed. Rhapsodizing about its elegance he wrote, "Now it twists about as if it would tie its body in a love-knot; then hangs motionless in some one of the 'lines of beauty' in which it has accidentally paused." Its eyes were especially gorgeous, he said; "the pupil is encircled by a fine ring of golden red, and the iris is marked with alternate divergent bands of grey and brown." Sadly, during its internment, the little pipefish succumbed to a disease that Gosse compared to dropsy in humans, known these days as edema.[19] He tried to cure his syngnathid captive by gently popping the air pockets that bubbled up under its skin, but he came home one day to find his pet dead. Then, in a later book, having shown such devotion to the ailing creature, he dismissed pipefish as being "rather uninteresting tenants of an aquarium."

As the frenzy for public and private aquariums spread, so did a fixation with seahorse watching. The first seahorses to go on display to the public were at Gosse's London Fish House. In 1859, a Mr. Pinto brought back four live seahorses from the mouth of the river Tagus in Portugal, transporting them carefully in a glass bowl of seawater. Pinto endured a virtually sleepless, seven-day train journey through Europe, rousing himself frequently to aerate the seahorses' traveling water with a syringe.[20] As soon as they were installed in their new London home, the seahorses were an instant hit. For many people, the seahorses had swum straight out of the pages of fairy-tale storybooks and into real life; they had to be seen to be believed. Before long, seahorses had planted themselves firmly in Victorian imaginations, with visitors flocking to see these beguiling creatures in aquariums all around Europe. In 1866, the *Glasgow Herald* lovingly described the new hippocampus arrivals at the Jardin d'Acclimatation in Paris. "They will entwine their tails together, and if a third or fourth should happen to join them the latter's tails will get entangled with those of the former, and thus a sort of nose-gay is formed, which will swim to and fro according to the will of the strongest, until it pleases them to separate again."[21] By 1873, one newspaper reported that seahorses had become "popular among naturalists" and "the hero of the aquarium, and the pet of the London ladies, who cultivate monsters with so much gusto in their back drawing-rooms."²²

Seahorses at the Brussels Zoological Gardens were featured in the *Daily Telegraph* in 1869, with their "prickly manes, and a motor in the idle of their backs that strongly resembles the fly wheel of a musical box."²³ At the Manchester Aquarium it was said the "sight of the antics of those beautiful creatures the sea horses is quite a treat, and alone worth a visit." It was within these glassy walls in the British Midlands that a herd of baby seahorses was born in 1873, possibly the first seahorses to start life in a public aquarium. One came into the world at one o'clock in the afternoon, followed by twelve more the next day. A journalist from the *Manchester Courier* remarked on the energetic nature of the newborns, which weren't sitting around quietly like adults but "moving about in every direction as if in search of food."²⁴ Seahorses were top of the bill at the grand openings of the Great Yarmouth Aquarium in September 1876 and the Scarborough Aquarium in May 1877. The Royal Zoological Society of Ireland, Hamburg Aquarium, Brighton Aquarium, Alexandra Park, Crystal Palace, and the Royal Aquarium at Westminster all played host to seahorses for the entertainment of Victorian visitors.

These "queer fish" also became a favorite topic for newspapers and magazines, where writers found great amusement and much amazement in the seahorses' odd looks and strange behavior. "Most people have heard of, and many have seen, either alive or dead, the remarkable specimen of the finny tribe," wrote Reginald A. R. Bennett in *Boys Own Paper* in 1898.²⁵ "See how proud Mr. Hip. appears to be of Mrs. Ditto," wrote Uncle James in a 1869 article "Wonders of the Deep." "They have no hands to shake each others fists with, and so they shake tails."²⁶ A story in *Little Folks* describes the author's visit to a Parisian store filled with glass tanks of fish, sea anemones, and a troop of seahorses "enough to mount a whole regiment of sea fairies."[27] Even the seahorses' ornate, antipodean relatives, the seadragons, were written about. In the *Ladies' Treasury: A Household Magazine* in 1892, one writer mistakenly described a drawing of a sea-dragon as a seahorse that had adorned itself in ribbons of sea-weed stuck on with a coating of gummy mucus. "This decoration falls off as the fish gets into deep water," we are told, resulting in the more familiar form that seahorses take in aquariums.[28] In 1875, the comic *Funny Folks* featured seahorses in a poetic reply to a local magistrate's refusal to grant the Westminster aquar-ium a license allowing its clientele to dance:

> *Within the Royal Aquarium*
> *What monsters have they prancing,*
> *That the authorities refuse*
> *A license there for dancing?*
>
> *Are they afraid the octopus*
> *Will writhe in a fandango?*
> *Or lobsters caper and pousette*
> *To airs from "Madame Angot"?*
>
> *Will lively crabs the can-can dance*
> *And learn to bounce and royster?*
> *Or flounders wobble to the strains*
> *Accomplished whistling oyster?*
>
> *Are seals as ready for a spin,*
> *As Gray says, when (look o-er him)*
> *"My grave Lord-Keeper led the brawls,*
> *The seals" they "danced before him."*

Will uncrimpt skates, as in a rink,
Be dreadfully disporting?
And hippocampus-major take
To capering and snorting?[29]

Having reached such notorious heights as a fashionable fad, nature began to lose its hold on the general public's interest, and so, almost inevitably, the bubble of Victorian home aquariums burst. People discovered that keeping sea creatures at home was more troublesome than Gosse and others had professed. "Balancing" an aquarium was not so easily done and often resulted in stinking chaos. In upper-class households, maids were left to clear up, resentful when their daily tasks involved chasing runaway crabs and errant pond snails around the house. The butler in Wilkie Collins's 1868 novel *The Moonstone* was no doubt grumbling about aquariums and other natural historical crazes when he said, "Nine times out of ten they take to torturing something, or to spoiling something and they firmly believe they are improving their minds, when the plain truth is, they are only making a mess in the house."[30]

But this was not, of course, the end of the aquarium, merely a cooling-off period from a feverish aquarium mania, which continued to simmer gently away. Today there are hundreds of public aquariums around the world ranging from small collections of local critters to glossy oceanariums housing giant whale sharks, shoals of tuna, and performing killer whales. Naturally, much has changed since the time of Philip Henry Gosse, changes that had already been dreamed up in the minds of many Victorian naturalists. The *Hand Book of the Fish House*, a slim volume from 1860 sold to visitors at Regent's Park, forecast that "when the mode of transporting delicate creatures from a distance is better understood, tropical seas, and even coral reefs, may contribute to our stores and enable us successfully to carry out the idea of a true Water-Menagerie."[31] H. Noel Humphreys made similar predictions in *Ocean Gardens*, where he labeled the marine aquarium "as yet, a plaything, a mere toy; but it is destined to become a far more important means of advancing science." He proclaimed that "we are now entitled to expect from science, that it shall exhibit to us the wonders of the tropic deeps" and one day display "the glorious shells of those regions" and that "fish gleaming with unusual dyes—metallic azure, and silvery crimson—will dart in our tropical tempered tanks as in their own tropic ocean, for our delight and gratification."[32] And in imagining the scientific advances and exotic creatures we would eventually all gaze upon, Humphreys was absolutely right.

What would Philip Henry Gosse have thought if he had been allowed a visit to an aquarium a hundred and fifty years after he first put sea creatures on public display? Perhaps we could take him to one of the many aquariums that specialize in exhibiting seahorses and the rest of the Syngnathidae family. Gosse could stand on Walnut Street Bridge in Chattanooga, Tennessee, and gaze across at the giant glass pyramids of the aquarium, reminding him perhaps of the glass palaces of Victorian London, only so much bigger and more angular. Through the smooth parting of the glass entrance doors, he would take his first ride down an escalator and descend into a dark hallway lit by glowing signs welcoming him to "Seahorses: Beyond Imagination."[33] He would hover for a moment, puzzling at the moving montage of seahorses on a bank of television screens before following the murmurs of the crowd into a room full of seahorses.

A familiar atmosphere of wonderment unfolds among the daily crowds at the Tennessee Aquarium, just as it must have done many years ago at the London Fish House. Kids run frenetically from tank to tank, drowning out gentle aquarium melodies with cries of "I wanna ride a seahorse," "Look how fat they are!" "Seahorse! Seahorse! Look, Mom!" "They look sad," "These ones have wings like dragons. I'll call them dragon horses!" Some children press their faces close to the glass tank walls, trying to spot tiny dwarf seahorses, each one as tall as a postage stamp but perfectly formed, hiding between blades of Florida Keys seagrass. "I can't even see them," "I found it I found it!" "Look how teeny tiny they are." Then the children take turns to pop up inside the hollow dome of a large cylindrical tank and, surrounded by delicate seahorses, imagine they are sitting on the seafloor. Meanwhile, adult visitors with no kids to herd stand quietly, smiling and still, intently watching the slow-moving seahorses, whispering to each other, "I've never seen anything like it in my life." Couples hold hands and point knowingly at the seahorse lovers holding tails.

Inside the illuminated tanks are convincing resin replicas of the seahorses' wild habitats, a submerged forest of mangrove roots, a shelly bivalve reef from Chesapeake Bay, the shady, green canopy of a Tasmanian kelp forest, and the branching corals of a tropical reef. In this imitation ocean world live not just one type of seahorse like the Victorians had, but seven: yellow, dwarf, pot-bellied, slender, tigersnout, lined and White's, as well as both of the world's seadragons, leafy and weedy, and a handful of different pipefish, alligators, banded, and blue striped. Bright display boards beside each tank provide a snapshot of what scientists have learned in a century of biological advances since syngnathids were first shown to the public. They tell of the seahorses' untamed lives and warn of the problems they face in the modern world: habitat destruction, overfishing, climate change, and coral bleaching—all words the Victorian naturalists would not have known.

Philip Henry Gosse would no doubt have been awestruck to

discover how far his idea of public aquariums has advanced and been delighted at how popular they remain today. After admiring seahorses in Chattanooga, he would surely have been eager to find out how the gleaming displays are created and, checking that no one was watching, he would take his chance and slip through the Staff Only door. Behind the scenes is a dazzling tungsten-lit world, a jungle of bubbling tanks and whirring valves, pipes, sumps, pumps, and siphons: all the necessary life-support systems to keep so many fish alive so far from the ocean. On one wall, a row of taps labeled WARM SALT WATER and COLD SALT WATER connect to giant storage tanks of "Instant Ocean," a twenty-first-century formula of salts to make artificial seawater, similar to one Gosse himself invented. Water from each display tank is dripped through a large column of plastic pebbles where bacteria break down ammonia and prevent it from accumulating to toxic levels in the fish-filled water. Another gadget bombards water with UV rays, powerful enough to destroy virtually all unwelcome diseases and parasites. One key secret to the crystal-clear water is a device called a protein skimmer that churns a barrel with artificial waves, separating and skimming off the same frothy, fatty scum that marks high tide on a beach. And every day, the aquarium keepers climb up a ladder and reach into each tall tank with a special underwater vacuum cleaner that slurps muck off individual stones and pebbles.[34]

Deeper into the concrete underbelly of the aquarium is a room lined with row upon row of tanks wriggling with new baby seahorses. The Tennessee Aquarium, like many others, keeps male and female seahorses together in the same public display tanks. And with no particular encouragement needed, this nearly always results in the arrival of more seahorses. Each morning, before visitors are allowed in, the aquarium lights are raised slowly and the seahorses, roused from their dormant nighttime state, greet each other, dance, flirt, and mate. At any one time, a number of the males on display will proudly sport a bulging pregnant bump. And each morning, aquarium keepers check the tanks and siphon off any new arrivals, tenderly taking them away to be reared in playpens behind the scenes.

Inside the rows of small tanks, minute transparent specks of day-old seahorses bob and gyrate, dancing to their own silent rhythms. Oddly, the very tiniest babies come from one of the largest seahorses, the slender seahorse, though in great abundance, up to sixteen hundred at a time. These diminutive newborns take a while to learn how to use their tails for grabbing hold of things and so, for the first few weeks of life, they have to be kept moving in a gentle whirlpool because otherwise they would drift down and lie helpless on the tank floor. And being so tiny, the slender seahorse youngsters come equipped with snouts so narrow they can suck on only the smallest of food. Across, in one corner of the seahorse nursery, sit neon-green bubbling jars filled with a thick soup of single-celled algae or phytoplankton. The algae are fed, a cup at a time, to a tank of minute animals called rotifers that grow to the size of a grain of sand, just small enough for baby seahorses to nibble on. Rotifers are also fed to other animals a shade bigger called brine shrimp, otherwise known as Sea-Monkeys, a popular breed of microscopic pet. Brine shrimp are fed to slightly older seahorses as well as other microscopic creatures called mysid shrimp, the staple diet of adult seahorses both in captivity and in the wild. Thus, step-by-step, the aquarium keepers create their own food chain, mimicking the complicated food webs found in nature to provide food of many sizes to keep generations of seahorses well fed. Eventually, though, most adults are weaned onto frozen mysids bought cheaply in bulk.[35]

One thing missing from the syngnathid nursery wards are infant seadragons. Compared to seahorses and pipefish, seadragons are notoriously and mysteriously difficult to breed in aquariums. Leafy seadragons have never successfully bred in captivity; any that you see in aquariums will have come from the wild. They often try to mate, sometimes quite single-mindedly. They dance together each day, the female desperately attempting to stick eggs onto her partner's belly (seadragons don't have enclosed pouches), but so far it has not worked. One pair at the Tennessee Aquarium courted day in, day out, for a whole year, until one morning, keepers found the male had dropped down dead, exhausted from his gargantuan but unproductive efforts. Male weedy seadragons have occasionally reared clutches of offspring, but for some reason, they find it much more difficult than seahorses. In the wild, seadragons live in relatively deep waters, so aquarists are experimenting with keeping them in five-meter-tall tanks in the hope it will give these resplendent creatures enough room to perform their pre-mating rituals.[36]

Around the world, there are about eighteen seahorse species that now live in aquariums, depending on exactly which ones you accept as being true species. Most of them will breed happily in captivity, but so far only eleven species have been raised to adulthood, with so-called closed life cycles. The two European seahorses, the long-snouted and short-snouted seahorses, have proved especially tricky to rear in captivity and no one really knows why. When breeding is successful, aquariums will often end up with a glut of seahorses on their hands; these are packed up and sent to other aquariums, creating a captive global population. The Birch Aquarium in La Jolla, California, first began specializing in local Pacific seahorses in 1994 and now raises nine different species, sending hundreds around the world every year. This means fewer seahorses need to be taken from the wild each year to fill aquarium displays.[37]

The sight of delicate seahorses and seadragons going about their quiet lives in these public see-through worlds no doubt gives many aquarium visitors the idea of keeping their own syngnathid pets. There are thought to be around two million people who keep marine aquariums in their homes, and more and more are adding seahorses to their tanks. As a young boy, the great naturalist and avid animal collector Gerald Durrell succumbed to the temptation of keeping seahorses. In his book *Birds, Beasts and Relatives*, the follow-up to his famous *My Family and Other Animals*, Durrell wrote more stories from his childhood days living on the Mediterranean island of Corfu in the 1930s. On one particular occasion, a local fisherman friend called Kokino helped Durrell sift through stones and seaweed at the bottom of his nets for any treasures snagged from the seafloor. "Suddenly," Durrell wrote, "Kokino gave a little grunt of half surprise and half amusement, picked something out of a tangled skein of sea-weed and held it out to me on a calloused palm of his hand. I could hardly believe my eyes, for it was a sea horse. Browny green, carefully jointed, looking like some weird chess man, it lay on Kokino's hand, its strange protruding mouth gasping and its tail coiling and uncoiling frantically."[38]

Diligently searching through the rest of the pile, Durrell found six more seahorses. The rest of the afternoon was spent eagerly dashing back and forth from villa to sea and back, first to evict a family of slow worms, making space in their tank for the new additions, then to collect bucket after bucket of seawater. Eventually, the seahorses themselves were installed in their new home in Durrell's bedroom and "like a group of ponies freshly released in a field, they sped round and round the aquarium, their fins moving so fast that you could not see them and each one gave the appearance of being driven by some small internal motor."

The seahorses were a great hit with the rest of the Durrell family, who all came to watch them "zooming and bobbing to and fro in their tank." But he rapidly faced the age-old problem of keeping sea creatures alive in a small tank. Without a system of aeration or a source of microscopic food, Durrell soon became exhausted from trekking to the sea four or five times each day to collect fresh, food-rich water for his seahorses. On the verge of giving up, one of the seahorses gave birth and Durrell vowed to hold on to the tiny swarm for a few more days until his mentor, Theodore Stephanides, was due to visit. It was only then that Durrell was told the truth about seahorse sex lives. "What I had thought was a proud mother was in reality a proud father."

Keeping fish is an inherently melancholic pursuit. There are no long walks to be had and no fetching of sticks, no playing with balls of wool or soft fur to stroke. Tank inhabitants can undeniably put on a splendid display but the pet-owner relationship is strictly one-way: owner to fish. Keeping seahorses, though, seems to be a little different. There surely can be no pet more magical than a seahorse, a real-life unicorn in your own home. And these are pets you can apparently get to know and pets that will get to know you. Seahorse owners are convinced that each one has its own personality. Some are show-offs, some are shy, some are affectionate, some put on airs and graces, politely waiting at meal times for their tank mates to join them rather than eating alone. Some are smart, some are lazy, some are head-butting bullies, and some are cheeky, snicking playfully at each other. Many seahorses are trusting, hooking on to their keepers' fingers while their tanks are cleaned; some are clingy, refusing food from anyone but their owners.

Such is the appeal of these fishy characters that seahorse keeping has become a twenty-first-century craze, with many contemporary collectors matching the devotion of the most ardent Victorian naturalists. Virtual communities of seahorse keepers from all around the world meet up online in chat rooms where they exchange stories and tips, put baby seahorses up for adoption, show photographs, and ask questions. "Is my male sterile?" "Why won't they mate?" "Am I overfeeding them?" "What do I do with my new arrivals?" Keepers give their pet seahorses names like Poseidon, Triton, Thrasher, Pacer, Charlotte, Sea Biscuit, Mrs. Speckles, and Fat Albert; they speak to them, sing to them, and cry when they die. And in return, seahorses take over lives. Despite all the modern advances in aquarium keeping, seahorses are still a lot of hard work and require an unrelenting commitment. Devoted keepers can't bear to be away from their seahorses who, like Goldilocks, must have things "just right": water temperature, light levels, acidity, nutrient levels, and, of course, food. Vacations are forgone so keepers can be on hand to feed their pets. Seahorse stomachs are small and don't hold much food, which means they are almost always hungry. In the wild, when they are not sleeping, flirting, or mating, seahorses spend most of their time eating. If they were allowed to, they would do the same in captivity, but they have to make do with two or three times a day. And their limited digestion also means that much of what seahorses eat passes straight through, and their tanks need a lot of cleaning.[39]

As Philip Henry Gosse discovered, captive syngathids can be highly sensitive and are prone to a host of ailments and diseases. It is important to keep a close eye on seahorse pets and watch out for any signs of illness: labored breathing, "coughing" of gills, blocked snouts, limp, drooping fins, or itchy skin infections that need scratching on tank walls. A collapsed air bladder leaves seahorses negatively buoyant and floundering on the tank floor, while males can get air bubbles lodged in their pouches, causing them to bob hopelessly at the water surface like Ping-Pong balls.[40] Seahorse owners will often keep an emergency hospital tank on standby, so any sick horses can be quickly isolated and bathed in water-soluble drugs. Syngnathids have, however, occasionally proved to be surprisingly hardy. In the aftermath of 2005's Hurricane Katrina, the emergency generator at the Audubon Aquarium of the Americas in New Orleans failed, and as the air pumps ground to a halt, nearly all of the ten thousand fish suffocated. Among the few survivors were the weedy and leafy seadragons, which somehow managed to hold on.[41]

As for where these seahorse pets come from, you won't spot modern-day aquarium keepers scrambling on the shoreline, scraping, scooping, and plucking their exhibits from the sea like their Victorian predecessors. Instead, at the click of a mouse, all you ever needed to keep an aquarium will be on its way to your door, including, for many seahorse enthusiasts, pets that have never seen the wild but are reared in a growing number of seahorse farms.

The earliest attempts to rear seahorses in captivity were in China in the 1950s, when people imagined good money could be made supplying the trade in traditional medicines. Back then, not enough was known about how to keep these finicky fish happy away from the oceans, and the farms never got off the ground. It wasn't until forty years later that seahorse husbandry was finally cracked and couples could be made comfortable enough in aquarium tanks to court, mate, and rear offspring.[42] The original motivation for the first successful seahorse farm was also to supply medicine markets, but not purely to make money. The hope was that farming seahorses would alleviate pressure on wild populations. But a pilot project at an aquaculture institute at the University of Tasmania in Australia quickly revealed that the numbers would not add up.[43] It was impossible to rear seahorses cheaply enough to be competitive in a global market where dried seahorses are sold by the ton. Soon, however, another set of customers emerged who were prepared to pay high prices for good-quality live seahorses. Aquarium keepers, lured by the idea of keeping their own enchanting seahorses at home, became the new customers queuing up for farmed specimens.

Today, there are a handful of seahorse farms established around the world, where regimented seahorse couples are kept in pristine isolation for the sole purpose of producing seahorses for aquariums. The seahorse production line usually begins with "Adam and Eve" adults that are taken from the wild, paired up, and placed in a tank. Nervous farmers watch on, crossing their fingers that nature will take its course and the couples will soon begin dancing. Once the young ones arrive, they are whisked away to a seahorse nursery ward. A few captive babies will be reared to adulthood and kept on to help produce the next generation of seahorses; most will be kept for just six months before they are ready to be packed up for export, often sold (at a premium) as pre-paired male-female couples. Tied up safely inside plastic bags filled with oxygen-rich water and a little perch to hold on to, seahorses can survive surprisingly well on trans-global cargo flights. Seahorses have been international aeronauts since the 1920s, when hundreds of specimens for the London Zoo's aquarium were flown in from Paris, after a train ride from the Atlantic coast.[44] More recently, a consignment of unfortunate seahorses even endured a journey via normal airmail. In 2006, customs officers at England's Stansted Airport seized a cardboard box containing one hundred dwarf seahorses that were sent by regular parcel post from Florida.[45] Miraculously, ninety-two of them survived.

Many different species of farmed seahorse are now available and are often given new commercial names such as White Knights, Southern Champions, Chargers, and Asian Emperors. One of the largest seahorse farms, Ocean Rider based in Hawaii, charges top dollar for their patented, pedigree seahorses, bred for their vibrant colors and sold under trademarked names: Mustangs, Zulu-Lulus, Fire Reds, and Sunburst. A recent addition to their collection is a new breed, the Pinto seahorse. The Ocean Rider Web site doesn't reveal what species or hybrid they are, simply describing them as "the most desirable and unique seahorses ever!" and at nine hundred dollars each, they must be the world's priciest syngnathids.

Seahorse farmers are careful to point out their environmental and ethical credentials. They urge customers to consider buying seahorses only when they are already well-experienced aquarium keepers. There is little point in dropping a few expensive seahorses into a tank if they are just going to wither away. Ocean Rider refuses to sell their seahorses in Hawaii, fearing for the endemic seahorse population should these exotic cousins be accidentally released.[46] And most farms work closely with CITES, demonstrating the minimal environmental impact of their operations in order to gain the necessary international export permits. But, nonetheless, much debate remains over whether the boom in seahorse farming is truly helping to reduce pressure on wild populations.

Even if a more cost-effective method of farming seahorses was one day to come along, there is still little chance these captive-bred seahorses would ever be used for traditional medicines, which make up the bulk of the global seahorse trade.[47] In the past, many attempts have failed to divert the attention of lucrative markets away from endangered species by farming them. From the point of view of simple supply-side economics, it would seem easy to flood the market with farmed specimens, drive down prices, and eliminate incentives to take them from the wild. In reality, this has rarely happened. The "tame" farmed species are often deemed inferior to wild-caught equivalents. Wild tastes better, wild is more potent. Wild is often, therefore, highly valuable and so the wild trade invariably continues.[48] Given their use in herbal medicines that promote virility and male potency, tamer, farmed seahorses will never be as popular as those taken from the wild. If one day cheap, farmed seahorses were to emerge on to the world market, it is possible they could give rise to a new generation of everyday seahorse products. That is now the fate of farmed bears that are exploited as the source of another Chinese medicinal ingredient. Bear bile was traditionally extracted from the gall bladders of slaughtered wild bears until methods were developed to "milk" live captive bears. The farmed bile is extremely cheap—not to mention extremely cruel—to produce and now finds its way into many low-cost domestic products like shampoo and soap while demand for "real" wild bile for medicines continues.[49]

Fortunately, when it comes to trading seahorses as pets, farmed specimens are often preferred. Before seahorse farms came along, few people kept pet seahorses. Enticing as they may be, seahorses taken from the wild and put in an aquarium rarely adapt well to captive conditions. They get stressed and easily succumb to diseases brought with them from the ocean, and they starve themselves unless presented with the freshest of squirming live food. Conscientious farmers wean their seahorses onto frozen food and make sure they are free from infections and diseases before selling them. So, despite their initial reluctance, aquarium keepers soon realized that captive-bred seahorses were healthier, were easier to look after, and lived much longer in tanks than wild-caught varieties.

Seahorses for the pet trade are now mostly bred carefully in captivity, but still the system isn't perfect. Even though farmed seahorses are clearly a better choice for aquariums, there remains the problem that trade routes for farmed seahorses open up laundering opportunities for wild-caught specimens. Farmed seahorses commonly get mixed in with others that are either wild-caught or "ranched," and consumers may not always know what they are getting. Seahorse ranches, unlike well-run seahorse farms, still ultimately rely on wild stocks. Pregnant males are taken from the wild and held captive in birthing tanks. Their resulting offspring are then reared and sent off to the aquarium trade. More pregnant males are brought in, and the next batch of young seahorses is processed. Arguably, this treads more lightly on wild populations than direct capture. During their first few months and years at sea, many young seahorses will die, and chances of survival are much higher in captivity. However, the captured males make no contribution to the next generation of seahorses, and their departure from the open sea upsets mating cycles of the female partners left behind. Ranching also tends to upset seahorse farmers, who are working hard to establish an ethical, low-impact industry. Ocean Rider is the first seahorse farm to develop a random tagging method to distinguish their seahorses in the marketplace.

One possibility to help boost dwindling populations could be to release captive-bred seahorses into the wild. Restocking the seas is certainly a worthy idea, but as with the liberation of any captive animal, it is tinged with controversy. There are those who think conservation programs should direct their attention toward protecting wild species and their natural environment instead of bringing them on board a protective, artificial ark. What point is there, they argue, in breeding endangered animals in captivity if their wild ecosystems continue to vanish? Shouldn't valuable conservation dollars be spent on securing the future of natural habitat rather than coaxing a few individuals to mate within the confines of a zoo or aquarium? Whatever your viewpoint on expensive ex-situ breeding programs, there is no denying that several species such as black-footed ferrets, Arabian oryx, and American bison might no longer be walking the earth if captive breeders hadn't given them a helping hand.

At least seahorses are among the cheaper species to breed in captivity. They reproduce abundantly, grow quickly, and take up much less room compared to tigers or rhinos, and there are few of the behavioral problems that other species exhibit. Many captive-breeding programs spend huge budgets on training naive animals to behave properly in the wild. Humans dressed in wetsuits and snorkels try to teach orphaned sea otters how to use stone tools to crack open spiny sea urchins, whooping cranes are taught to migrate a thousand miles from Wisconsin to Florida by solo human pilots in micro-light aircraft leading the way, and male giant pandas are shown films on how to woo members of the opposite sex. Most farmed seahorses are inherently skilled in mating and hunting and generally staying alive. But this does not necessarily mean that all attempts to breed seahorses in captivity will be an unfailing success. Often, what we think of as two separate seahorse species will in fact mate quite happily when kept together in an aquarium tank. With such fragile and fuzzy species boundaries, who's to say that, given the opportunity, the same wouldn't happen in the wild? If a seahorse species is released into an area where it doesn't normally live, there is a chance it will hybridize with native species and disrupt the gene pool. Another problem is, when small populations breed in captivity, they tend to become inbred and will be less well suited if they are eventually put back in their natural environment. Captive animals don't have the boisterous forces of nature to contend with, and weaker individuals tend to survive that otherwise would not cope.[50] Farmed seahorses released into the sea may bring with them sloppy genes making them less fit for survival in the wild.

But all this doesn't necessarily mean that seahorse restocking is impossible or misguided. A pilot scheme in Australia has recently released seahorses into some of the busiest waters in the country.[51] The liberated seahorses were White's seahorses, one of the many species described by Dutch naturalist Pieter Bleeker in the eighteenth century. He named them after John White, the first surgeon general sent to the then-British colony of New South Wales in the 1780s.[52] White didn't much like Australia but was passionate about the wildlife he found there, describing many new species in his *Journal of a Voyage to New South Wales.* He assumed the seahorse he saw there was the same one Linnaeus had recently described from Europe, writing, "This animal, like the Flying Fish, being commonly known, a description is not necessary."[53] Seventy years later, Bleeker thought differently and gave them their own name.

White's seahorses live on seagrass beds and sponge gardens in a corner of southeast Australia, between Sydney and the small town of Forster three hundred kilometers to the north. Like many seahorses, they have taken a liking to man-made structures, including the nets set up at municipal beaches around Sydney Harbour to cordon off safe swimming areas from boat traffic and protect swimmers from sharks. Not long after they are fixed in place, the nets become engulfed by sea life and provide a perfect home for the White's, which like to hook their tails on the ropes and hide among sponges and seaweeds. But when the burden of encrusting plant and animal life becomes too heavy, the nets collapse and local councils periodically remove them, taking the resident seahorse populations with them. Inspired by the seahorses' liking for these man-made nets, researcher David Harasti came up with a novel way to help encourage the dwindling seahorse population. In 2007, based on an idea similar to commercial seahorse ranches, several adult White's seahorses were carefully removed from Manly Cove, a northern suburb of Sydney, and transferred to a tank in Sydney Aquarium. There they paired up, danced together, mated, and eventually produced a shoal of new baby seahorses. But instead of being sold off to aquarium keepers, when they were four months old these seahorses were taken back and released onto the Manly Cove swimming net. The hope was that by drastically improving the survival rate of newborn seahorses, the net population would bounce back, and since only first-generation wild seahorses were used, there was no risk of genetic contamination.

So far, the experiment has been a success, with seven out of thirty released seahorses still seen regularly clinging to the nets six months later; it is hoped the others simply wandered off into nearby seagrasses. Harasti also discovered that the seahorses tend to hang out at the bottom of the nets, so an obvious alternative to replacing whole nets in the future is for divers to clean off the top sections, leaving growth lower down for the seahorses to nestle in. When all is said and done, it wouldn't be realistic to contemplate restocking large areas of sea with seahorses, but small-scale projects like this can play an important part in raising peoples' awareness of seahorses. Until the releases at Manly Cove, local residents didn't know they had seahorses living on their doorstep. Now these tiny animals have become a source of great local pride.

A century and a half after the first marine aquariums were set up, nature is no longer in vogue as it once was. Our lives are now crowded with a thousand other man-made distractions. While there are still naturalists and bird-watchers in our midst, we are more likely to memorize song lyrics or the names of football players than the trees in a local park or the wild animals that scamper through our gardens. Many of us do, at least, still share some of the Victorian sense of awe and wonder in nature, only instead of going out to find it in our backyards or on a trip to the beach, we turn on our televisions and switch on our computers to tune into the rest of the world. Nature documentaries on television mean we can all now watch the sunrise over a rain forest canopy and hear the shrill shrieks of howler monkeys calling each other awake. We can contemplate the darkness of the deepest ocean abyss, follow the wandering albatross on thousand-mile journeys, and crawl along the underground tunnels of burrowing naked mole rats. Victorian naturalists would have been dumbfounded by the amount of information about the natural world that lies effortlessly at our fingertips.

But there is one thing the Victorians had that we have lost forever. Back then, it was innocently and perhaps naively believed that the beauty of nature was never-ending and untouchable. Naturalists believed they could help themselves to as much of it as they wanted with no harm done. Admittedly, it wasn't long before they began to notice changes. Philip Henry Gosse himself was perhaps the first to suspect that things were not quite what they used to be. In 1856, he wrote to the aquarium supplier Lloyd, complaining that he now had to venture farther from his usual collecting spots to find tide pools stocked with the variety and abundance of sea anemones that he expected. He waved a finger of blame at either the winter frosts of the previous years or the "capacity of amateurs" for "extirpating the Actiniae (sea anemones) that were so abundant."[54] Nowadays, of course, not a day goes by when we don't hear another story of the emptying seas, the fragmenting rain forest, the dissolving coral reefs, or the collapsing ice sheets. We must all share the burden of regret, watching in vivid detail all the many ways that mankind is spoiling the wild world.

There are two emerging responses to these clarion calls announcing the perceived threat of planetary death. The depressing news acts as a cattle prod for some people, startling them into action, inspiring them to find ways to try and help. Then there is also a rising symptom of "green fatigue." It is easy to get exhausted and feel utterly powerless when faced with so much doom and gloom until, eventually, we throw up our hands, throw in the towel, and wonder what might have been if we hadn't let things go so wrong. In any case, with so much changing so far away, what real effect does it have on our lives anyway?

Maybe that is what makes well-run aquariums and zoos all the more important today. They remind us of the simple pleasure of standing and gazing at wonderful creatures in the flesh, right there in front of us. They let our imaginations wander off as we marvel at the glowing colors of coral reef fish, gasp at the tiny proportions and deadly displays of poison arrow frogs, get lost in an industrious colony of leaf-cutter ants, and admire the grace of a shark sliding through the water. These animals may not have their freedom, but they offer us a chance to reconnect with nature, lending meaning to the inflating inventory of ecological disasters going on in the wider world beyond the concrete pavements of the cities and towns where so many of us live.

There is certainly a lot we can learn from seahorses in captivity. They offer us a delicate glimpse of the wonders of the sea, they keep us puzzling over what else could be out there, and they give us an answer once and for all to the wavering question, "Are they real?" But we must remember that a seahorse in a tank does not equal a seahorse in the wild, a Bargibant's seahorse clinging to a sea fan on a deep Indonesian reef or a lined seahorse hiding in the shadows of a Belizean mangrove forest. Some may say that the seahorses kept in people's homes are no more than domesticated pets, as far removed from nature as a canary in a cage, a rabbit in a hutch, or a shih tzu on a leash. And yet it is hard to imagine there could be a single devoted seahorse keeper who doesn't sympathize with the need to protect the oceans, who doesn't also wonder how their pets' wild cousins are getting on. Even for those of us who have not chosen to bring live seahorses into our homes, there are other, numerous ways that seahorses quietly permeate our modern human world.

Chapter Six

WHY SEAHORSES MATTER

*E*ver since Aboriginal Australians painted Rainbow Serpents on cave walls, and ancient Celtic tribes carved enigmatic beasts in stone, the iconic shape of the seahorse has stayed with us. We may have solved many seahorse puzzles, found their ancestors and traced their evolution, still we celebrate seahorses in all their glory. They continue to provide great inspiration to artists, sculpturists, and jewelry makers. They take on a symbolic role as nautical emblems on heraldic coats of arms, coins, medals, and buildings. Instead of drawing them on tombs or coffins, we now decorate our bathrooms with them, on tiles, towels, sponges, and soaps. Seaside guesthouses, restaurants, and bars are named after them. And in Bahrain,

∞

Above: Hippocampus aterrimus, a synonym of the
yellow seahorse *Hippocampus kuda*
Credit: *Theodore Gill, 1905*

residents may one day soon live on a giant seahorse if plans go ahead to build an artificial island in the shape of the curly-tailed syngnathid.

Seahorses have also, for decades, infiltrated popular culture where they tell a host of different stories from the ocean. Often, the portrayal of modern-day seahorses simply reminds us that, unlike most other fishes, we consider real seahorses to be objects of beauty to gaze on, and not something to be eaten and exploited. In Wes Andersen's mischievous and melancholic movie *The Life Aquatic* from 2004, Steve Zissou—a thinly disguised homage to Jacques Cousteau played laconically by Bill Murray—is presented with a tiny rainbow-colored seahorse frolicking inside a plastic bag of water. "A Crayon Ponyfish," Zissou says. "Wow. Interesting specimen."[1] In Ruth M. Arthur's 1964 book *Carolina and the Sea-Horse*, young Carolina is on holiday in Venice when she sees a boy, Tomajo, selling dried dead seahorses. "Oh!" she cries, pummeling his arm with both fists. "How could you! You horrible boy! They should be alive and happily swimming about in the water," she says. "I cannot bear to see them shriveled and dead." Tomajo, whose fisherman father had died at sea, tries to explain to her that they are just fish, so why shouldn't he make money selling them to tourists? After showing Carolina live seahorses in a nearby rock pool, together they hatch a plan for Tomajo, when he starts his new job at the local glass factory, to make intricate glass seahorses worth much more than dead ones. Years later, Carolina receives a parcel containing a single perfect seahorse made of glass.[2]

The imaginary seahorse characters that appear in books and cartoons have adopted a range of personalities that perhaps reflect our various impressions of their real-world cousins. Sometimes, they are wild creatures that need to be tamed. In an offshoot of the Disney movie *The Little Mermaid*, the leading lady, Ariel, is determined to ride a feral seahorse called Stormy.[3] Despite being forbidden by her father, Ariel rides off into the wilderness, where she and Stormy get into a spot of trouble with a gang of seahorse rustlers.[4] A more ferocious interpretation of the wild seahorse character features in the Japanese computer game Pokémon, which became an international media franchise in the 1990s. The central idea of the game is for players to find and tame wild "pocket monsters," or Pokémon, and assemble a fighting team to do battle with other players and their monsters.[5] Out of nearly five hundred Pokémon,[6] there is a trio of characters called Horsea, Seadra, and Kingdra that are all based on members of the seahorse family. The Horsea are cute blue seahorses with yellow bellies and trumpet snouts. At certain points in the game, they can metamorphose into more powerful menacing monsters, the Seadra, with poisonous spiky wings. Like real seahorses, male Seadra raise the young and—less characteristically—will fiercely defend their nests from intruders. The Japanese practice of using real seahorses as traditional medicines has spilled over into this kid's game, with Seadra bones and fins being highly valued in the Pokémon universe as herbal remedies. The most fearsome of the seahorse pocket monsters, and most useful to have on your side, are the Kingdras, which evolve from Seadras and look like angry seadragons. They lurk in deep-sea caves, emerging to create stormy havoc, whipping up whirlpools and tornadoes with a flick of their formidable tails and unleashing maelstroms when they yawn.[7]

Several other seahorse cartoon characters are more lovable and fun. An adorable but greedy seahorse wins the heart of SpongeBob SquarePants in a 2002 episode of the Nickelodeon cartoon series created by animator and former marine biologist Stephen Hillenburg and set in a surreal underwater city called Bikini Bottom.[8] The seahorse in the Disney cartoon *Finding*

Nemo was a bit wet behind the gills and spent the duration of the movie sneezing because of his pitiful allergy to water.[9] And a clumsy seahorse appears in DreamWorks' movie *Shark Tale* and gets the main character into hot water after a day at the seahorse races.[10]

And sometimes, seahorses appear in our world today just because we like the idea of them. For a short while in the 1990s, there was a British indie band called The Seahorses. Led by ex–Stone Roses guitarist John Squire, their biggest hit was a catchy tune called "Love Is the Law," which went to number three in the UK singles chart in 1997 and reached number fifteen in the American charts. The British music magazine *NME* spread a rumor in 1996 that the band name was chosen as an anagram of "He Hates Roses," an idea dismissed by Squire in a later interview. They had in fact, he said, been considering the name for a while because it was a "dream symbol for travel and adventure." One night the band went for a drink at a club where there happened to be an enormous model seahorse. They'd been toying with the idea and took it as a sign. Some versions of the story even have Squire stumbling into it on his way to the toilet.[11]

Following the many thousands of years that we have brought seahorses into our human world—dead, alive, as images and ideas—there is little doubt that they still matter to us. They make us smile and laugh and they put dreamy thoughts in our heads about the exotic realm they live in. But we no longer have to make do with standing at the edges of the seahorses' world, peering in from afar and imagining what life must be like down in the inscrutable depths. Now we can join the seahorses and see for ourselves what goes on in their underwater kingdom.

Early deep-sea divers who walked across the seafloor with upturned buckets of air over their heads, like miniature mobile

diving bells, were the first to use technology to begin dismantling the physical barrier between people and the oceans. By the
nineteenth century, aquanauts were sealing themselves inside
rubber suits and breathing air pumped down to them through
a tube trailing from the surface. For a long time, it was only a
few rare, brave, and some might say mad pioneers of diving
who infiltrated the underwater world, but soon they began to
devise equipment for stealing snapshots of the oceans' secrets
and bringing them back for everyone else on dry land to see.
Underwater photography began in 1856 when William Thompson dangled a watertight camera on the end of a pole into the
chilly waters of Weymouth Harbour on England's southern
coast. His subject was a rather dull patch of seaweed. In 1893,
French scientist Louis Boutan took the first underwater photographs while diving. Then, in 1915, John Ernest Williamson
killed a shark in front of the world's first underwater movie camera. Not long after, moving pictures of wild seahorses were unveiled to the world when they made their film debut in one of
the earliest nature documentaries to be shot on location beneath the waves.[12]

Jean Painlevé was a pioneer of underwater filmmaking
twenty years before his fellow Frenchman, Jacques-Yves Cousteau, moored up *The Calypso* and leapt off the side to become
the world's most celebrated ocean adventurer. We all think of
Cousteau as the grandfather of underwater filmmaking, and few
have heard of Painlevé, whose name has slipped between the
pages of film history. Born in Paris in 1902, he was the son of
eminent mathematician and French prime minister Paul Painlevé. His life, like his filmmaking, was vivid, bustling, idiosyncratic, and eluded definition. For a while he followed in his
father's footsteps, studying mathematics, but he went on to
specialize in medicine and eventually marine biology at the

Sorbonne in Paris and published academic papers on the innermost details of marine creatures. As a young man, Painlevé became part of the flourishing avant-garde movement in Paris, he was a professional race car driver, and he was a founding member of the world's first dive club, named the Club Sous-L'Eau (a play on words, since *sous-l'eau* sounds a little like another French word, *soûler*, which would make it the drunkards' club). Later in life, he worked tirelessly to help foreign scientists and artists flee prewar France and eventually orchestrated his own underwater escape from the Nazi regime by diving underwater at night along a river flowing into Spain.

For a short time, Painlevé was a movie actor. It was while on set filming *L'inconnue des six jours*—his first and only appearance in front of the camera—that Painlevé was inspired to make his own motion picture. The cameraman, André Raymond, had worked out a way of creating a time-lapse effect by taking only one frame per turn of the camera's crank instead of sixteen. It was the beginning of a long collaboration between Painlevé and Raymond, and together they used this technique to film *The Stickleback's Egg: From Fertilisation to Hatching*. Shot through a microscope lens, the film tracks the development of fish embryos from the moment sperm and egg fuse, an astonishing cinematic first. But at its debut at the Académie des Sciences in Paris in 1928, *The Stickleback* received a deluge of scorn from an audience of outraged scientists. One botanist threw up his arms, declaring, "Cinema is not to be taken seriously, I'm leaving!" It was not the last time Painlevé would meet with such incredulous disbelief at his attempts to unite science and cinema. Undeterred, he went on to make over two hundred short films, some for a scientific audience, some for the general public. Each one focused on a different star of the sea—the octopus, the sea urchin, the hermit crab, the water flea, the lobster. They were shown before the main features at a new set of experimental ciné clubs that opened up around France in the 1920s and 1930s.

What remains remarkable about all of Painlevé's films to this day is his ability to seamlessly blend art with science. His work is rich with stylish imagery, set to stirring orchestral sound tracks and original jazz scores, and they were wittily and poetically narrated. The surrealist painter Marc Chagall admired Painlevé's film of shrimps and sea spiders as "genuine art, without fuss." And even though they are an exposition in aesthetics, his films do not shy away from scientific detail. Painlevé presented a gritty, close-up view of aquatic life, offering a new way of seeing and understanding nature, and at the same time he subtly held up a mirror to our own lives, pointing out similarities and differences of the human and aquatic worlds, effortlessly combining the familiar and beautiful with the unexpected and grotesque.[13]

His most famous and successful film was *L'Hippocampe ou Cheval Marin* ("The Seahorse"), released in 1934.[14] Painlevé explained his choice of subject as an attempt to reestablish the balance between male and female. "Everything about this animal, a victim of contradictory forces, suggests that it has disguised itself to escape, and in warding off the fiercest fates, it carries away the most diverse and unexpected possibilities. To those who struggle to improve their everyday luck, to those who wish for a companion who would forgo the usual selfishness in order to share their pains as well as their joys, this symbol of tenacity joins the most virile effort with the most maternal care."[15]

Filming for *L'Hippocampe* began in a Parisian basement, where Painlevé and Raymond set up a studio equipped with vast glass tanks draped in seaweed. The main protagonists of the film, pregnant male seahorses, had been brought in unceremoniously from the coast in rusting metal buckets. Once the seahorses with their stretched round bellies were installed in the miniature watery film set, Painlevé settled down to eagerly await the moment of birth. After several days of fruitless patience, Painlevé handed over the watch, returning several hours later only to find a sleeping Raymond and a tank full of baby seahorses. Determined not to miss the next male's labor, Painlevé constructed a device mounted on his hat that administered a small electric shock to wake him up whenever he nodded off. Such extreme measures paid off, and Painlevé finally got the footage he wanted.

The next scenes were shot in the Bay of Arcachon on France's Atlantic coast, where Painlevé crouched on the seabed a few meters down, armed with an enormous waterproof box, his camera peering out through a thick glass plate. Capturing wild seahorses on film for the first time was an arduous task. At that time, the only way Painlevé could spend time underwater was to gulp air through a rubber tube tethered to a hand pump at the surface. "What bothered me most," Painlevé later said, "was that at one point I was no longer getting any air. I rose hurriedly to the surface only to find the two seamen quarreling over the pace at which the wheel should be turned." To make matters worse, his camera held only a few seconds of film, requiring frequent trips to the surface to reload. But despite his submarine difficulties, Painlevé clearly enjoyed himself. "It was lovely: the underwater beauty seductive. It is easy to lose oneself in the water's depths."

The result of his efforts was a fifteen-minute black-and-white glimpse into the seahorses' world. Heralded by a trumpet fanfare, the film begins as an elegant drifting journey of intimately entwined tails. The audience is introduced to the demure seahorses; their upright stance gives them a pompous air, their pouting lips appear embarrassed, their mobile eyes anxious. "You can't help wanting to give this animal limbs or legs," the narrator tells us as we watch dozens of seahorses dance through festoons of sargassum. "When you see it moving about, its body vertical, its head horizontal . . . this aquatic vertebrate is strangely reminiscent of a biped."

Keen as always to tell the whole brutal story, Painlevé then shows us a dead seahorse sliced in half from head to tail, displaying its internal organs that are "like any other fish," the silver bauble of its inflated swim bladder shining in the camera lights like an aluminum foil balloon. The film then races headlong into the birthing scene, where with seismic contractions the pregnant male puffs, pants, and struggles his way through labor. Then we flinch as a scalpel leans in to cut open a dead male's belly, revealing a cluster of unborn babies imbedded in the honeycomb pouch wall fed by a tangle of blood vessels. We witness a live unborn seahorse with big round eyes, an upturned stubby snout like a King Charles spaniel, its belly attached to a giant yolk sac and already with a wriggling, tapering tail. Through its transparent body we watch in slow motion as blood pumps from chamber to chamber in the young developing heart. In extreme close-up, Painlevé lets us into the one of the seahorses' secrets. Clusters of dark cells shaped like dendritic ink spots are the chromatophores that swell up or shrink down, creating a coat of many colors. After this barrage of intense and intimate images, Painlevé leaves us to cool off with several uninterrupted minutes of peaceful seahorses swimming around their tank, once again the elegant, regal creatures we expected to see. Ever one to poke fun, he brings the film to a close with real seahorses in silhouette while a horse race plays out in the background, and then spells out the word *fin* in seahorses.

L'Hippocampe was Painlevé's only film to make any money, earning back the hundred thousand francs it had cost to produce. The seahorse blockbuster was the talk of France. Painlevé apparently overheard conversations on buses: "Did you see that movie about the male who gives birth?" Off the back of the film's success, he launched a line of seahorse jewelry that was sold in classy boutiques across France, displayed alongside aquarium tanks full of live seahorses. The range of necklaces, bracelets, earrings, and brooches were designed by his lifelong partner, Geneviève Hamon, and sold under the label JHP, for Jean Hippocampe Painlevé. But Painlevé and Hamon never saw any of the money they made selling jewelry. With little interest in the commercial side of the venture, Painlevé had handed over the running of the business to Clément Nauny, who, at the end of the war, made off with all the JHP profits and moved to Monaco, never to be heard from again.

Long gone are the days of the cumbersome diving gear that Painlevé used to capture the first moving images of seahorses. A decade after the release of *L'Hippocampe*, Émile Gagnan, a friend of Jacques Cousteau, was inspired by a design for a vapor-powered car to invent a system that delivered air to a diver from a portable pressurized air tank. It provided air whenever the diver inhaled and, more importantly, at just the right pressure at any depth underwater. Collaborating with Cousteau, he unveiled the Aqua-Lung, or the Self-Contained Underwater Breathing Apparatus—scuba—the piece of kit that has allowed people to finally achieve the dream of swimming free as the fishes. The advent of scuba diving has flung wide open the gates to the seahorses' underwater garden.[16]

Back in Victorian times, when the exploration of the oceans was still in its infancy, a German biologist, Johannes Walter, wondered whether newfangled innovations like aquariums and deep-sea submersible vessels would dismantle the beauty of the oceans, just as the invention of telescopes had—as some pessimists claimed—snatched the soul from the skies. What would Walter have thought of the easy access we now have to the oceans? If his despondent forecast was correct, then surely today with so many divers in the world—at least many millions[17]—there would not be an ounce of beauty left to find. But following the bubbles of Painlevé and Cousteau, many divers have proved Walter wrong, finding submerged magnificence in the aquatic realm, especially those lucky enough to have been blessed with a seahorse sighting. The joy of spying a wild seahorse is contagious. Any diver who has spotted a live seahorse will likely offer you a detailed account of the time they were diving in Indonesia or the Caribbean or wherever they were the first time they spied a miniature chessman of the sea. Either that or they coo and yelp that they would just love to see one but have never been in quite the right place at just the right time.

Despite the effortless safety of modern diving equipment, finding a seahorse in the wild is not easy. One can be right in front of your dive mask and still you won't see it as it hides beneath a shroud of crafty camouflage. The best thing is to stay in one small area and search as diligently as you can, watching patiently for the flutter of a tiny fin or the gentle grasp of a prehensile tail. Occasionally a seahorse will simply appear as if out of nowhere and casually swim past a row of wide-eyed divers. Seahorses live in virtually all the corners of the oceans—steering clear of the coldest climes—so, if you want to see one, there are many different places where you might get lucky. It helps that seahorses are generally not great travelers. Commonly, they stick to a small patch of habitat, which means once you know where one lives, you can visit it time after time and still find it clinging to the same coral branch, hooked onto the same bright orange sponge, or ambling about the same patch of seagrass. If you ask around, you might find someone, a fisherman, a snorkeler, or a beachcomber, who is happy to point you toward the resident seahorses, around the pilings of a jetty perhaps or in an area of mangrove where people don't often bother looking. Many dive centers around the world are starting to take great pride in their local seahorses and regard them as some of their main attractions. Dive guides can earn handsome tips when they take customers to see a seahorse and will often keep their valuable knowledge of seahorse hangouts a well-guarded secret from other guides. In the small island of Bonaire off the coast of Venezuela, there are at least twenty seahorse sites known to dive masters, and with at least twenty-five thousand vacationing divers on the island each year, these seahorses receive a lot of visitors.[18] In Indonesia's Bunaken National Marine Park, divers can sometimes see two or three different species of seahorse during a single dive.

Some tourism outfits have even begun to offer special seahorse-spotting tours. The lined and slender seahorses are just two of the many marine creatures that tourists flock to see in the shallow clear waters of Belize, the small Central American country nestled beneath the shoulder of Mexico's Yucatán Peninsula. The 386-kilometer coastline is fringed by the northern hemisphere's largest barrier reef and is punctuated by hundreds of low-lying coral and mangrove islets called cays. Seventeenth-century Caribbean pirates roamed the Belizean cays, illicitly setting up camps and launching attacks on Spanish galleons. They gave the islands names like Deadman's Cay, Drowned Cay, Rendezvous Cay, and Spanish Lookout Cay. Nowadays, some of the cays are privately owned by movie stars, and one has been converted into a luxury golf course, but many cater to tourists. At the northern end of the reef is Ambergris Cay, named after the whale oil that apparently used to wash up on the beaches. The colorful streets of the main town, San Pedro—immortalized by Madonna in her song "La Isla Bonita"—bustle with golf carts and vacationing families who take their pick from hundreds of diving, snorkeling, and sailing tours advertised along the beachfront. Among them, a few operators offer the chance to meet Belize's seahorses without even the need to get wet. In true James Bond style, tourists hold their breath as they skim across the flat backwaters of the cays in long skiff motorboats. The tour pulls up to a quiet spot where a guide hops out and peers carefully down through waist-high water, hunting around in the mangrove prop roots for some of the cay's tiniest inhabitants. There are never any absolute guarantees—nature is unpredictable after all—but more often than not, a seahorse is found. The guide lowers a glass jar into the water and gently scoops up the seahorse and for a few minutes the boatload of seahorse spotters can see for themselves the fish that they perhaps had once believed was nothing but a fairy tale.[19]

We might wonder whether such frequent contact with human admirers has any lasting effect on seahorses. Divers should obviously be respectful of the seahorses' quiet, sedentary lives and try to resist the urge to touch. While wrapped in a seahorse moment, it is easy to forget where your fins are and accidentally crash into nearby corals. But there is no doubt that divers can play an important role in proving that a seahorse is worth more in the water than out.

Keen seahorse spotters can also do their bit to provide practical help with scientific research, something that amateur stargazers and bird-watchers have done for centuries. Many thousands of pairs of divers' eyes scouring the sea can add up to a powerful seahorse task force that scientists have begun to tap into. Divers flock to Port Stephens, an inlet on Australia's east coast a few hours' drive north of Sydney, to see a rich mix of underwater species, some from cool temperate waters flowing up from the south and some from warm tropical waters in the north. Among the aquatic menagerie are White's seahorses, including seven hundred that have been tagged by seahorse researcher David Harasti. He has enlisted the help of local divers to track the seahorses, which he marks with tiny patterns of colored dots injected into their skin. Whenever divers spy a tagged seahorse, they take photographs and e-mail them to Harasti along with details of where and when it was seen. Tracking the same individuals provides vital data on how far seahorses move and how long they live. So far, they have shown themselves to be not especially energetic; one typical tagged male has not budged from his sea fan home for over a year.[20]

A similar but larger-scale project was set up in Italy called the Mediterranean Hippocampus Mission.[21] In 1999, researchers from the University of Bologna recruited and trained over two and a half thousand sport divers to conduct the first-ever survey of seahorses living along Italy's boot-shaped coastline. On "Hippocampus Day," workshops were run at dive centers across Italy to raise awareness for the campaign and provide instructors and dive guides with the necessary skills to teach other divers how to fill out seahorse-spotting questionnaires whenever they see, or don't see, seahorses while diving. Divers were taught how to distinguish between the two Mediterranean seahorses—the short-snouted seahorses tend to be quite bald compared to the long-snouted seahorses with their bristling long manes[22]—and prizes of magazine subscriptions and dive holidays were donated as incentives to get people involved. After three years and more than six thousand hours of accumulated time underwater, the volunteer divers spotted more than three thousand sea-

horses living all the way around Italy, in the Ligurian, Tirre-
nian, and Adriatic seas. Such a widespread census would have
been virtually impossible without the help of so many keen seahorse spotters; the same information would have taken a professional scientist twenty years to collect, at a cost of least a million
U.S. dollars.[23]

The long- and short-snouted seahorses are also the focus of
another amateur seahorse-spotting venture at the northern
limit of their European range. Since 1994, the British Seahorse
Trust has gathered reports of seahorse sightings from divers,
anglers, and beachcombers browsing the country's convoluted
shoreline. The original aim of the trust's director, Neil Garrick-
Maidment, was to find out once and for all whether seahorses
are permanent residents or merely fair-weather visitors to the
British Isles. As reports poured in, including underwater photographs and film of pregnant males, it became clear that both
the European seahorse species are indeed full-time members
of British marine fauna.[24] Another syngnathid project is Dragon
Search in Australia, which collects reports of seadragon sightings from members of the public, with individual dragons recognized from their distinctive facial patterns.[25]

The contribution divers make to seahorse studies doesn't
stop at simply pinpointing where they live. The devotion of one
particular seahorse diver and underwater photographer led to
the discovery of one of the smallest species of seahorse in the
world.

Denise Tackett and her husband, Larry, spent fourteen years
collecting sponges. Together they traveled the world from coral
reef to coral reef, camping on idyllic tropical beaches. Every day,
they would carefully gather sponges of all shapes, colors, and
sizes, then parcel them up and ship them back to America,

where researchers screened them for new chemicals to turn into the next generation of anticancer drugs. It was the perfect job for a couple who shared a great love of scuba diving and who dreamed of becoming successful underwater photographers. During their many thousands of dives, there was one creature that Denise wanted to see more than any other: a "pygmy" seahorse. The Bargibant's seahorse was the only known pygmy seahorse and it was without doubt the strangest-looking of them all. They were originally discovered by accident in 1969 by a scuba diver collecting specimens for the Nouméa Aquarium in New Caledonia. George Bargibant was hanging around at five meters waiting to "off-gas"—if divers ascend too quickly, they risk getting the bends, a potentially lethal condition caused by nitrogen bubbling out of the blood like a carbonated drink spurting from a shaken-up can—when he noticed a pair of most unusual creatures clinging on to a gorgonian sea fan he had just collected. He was sure they were seahorses, but they were unlike any he had seen before. They were bright cherry pink, had the shortest, stubbiest of snouts, and were covered in bulbous, white-headed nubbins that looked just ready to squeeze. He hadn't seen them at first because their pimply costumes blended in so perfectly with the unopened pink polyps of their sea fan home. And at just two and a half centimeters long, these were, at the time, the smallest seahorses in the world. Denise had seen photographs of those two original Bargibant's seahorses taken in the Nouméa Aquarium, and she longed to see a live one for herself. But no matter how hard she looked, the pygmy seahorses stayed hidden, like fairies at the bottom of the garden.

By the mid-1990s, Larry and Denise's sponge-collecting work had brought them to a stretch of water known as the Lembeh Straits on the northwestern tip of Sulawesi, an island that lies like a misshapen starfish in the middle of Indonesia. It was not long before they realized that this was a very special place. They found a staggering abundance of creatures, the likes of which they had not seen anywhere else on their tour of the world's tropical seas. And yet, still there were no pygmy sea-horses. That was until Denise met a pair of visiting underwater photographers who told her about their recent trip to Bali, where a dive guide had shown them tiny seahorses clinging to a type of sea fan called *Muricella*. This was the same fan, with splayed bony fingers filled in with a red lacy mesh, that George Bargibant had collected in New Caledonia twenty years previously. On discovering this vital clue to the pygmies' favorite hangout, Denise grabbed her dive gear and headed straight out to find some *Muricella*. And sure enough, there they were, ti-nier and more perfect than she had ever imagined. As soon as she started looking at the right type of sea fan, much finer and more delicate than the sea fans she had previously scoured, she saw pygmies everywhere. From then on, whenever Denise was out hunting for sponges, she always kept an eye open for *Muri-cella* sea fans and their delightful seahorse residents. She began building up a portfolio of photographs and became well known as quite the pygmy seahorse nerd.

The story continues a short while later, when Denise and Larry took a day off from sponge hunting and joined a dive trip to a nearby shipwreck. During the dive, one of the guides waved frantically at Denise and pointed toward a sea fan. It wasn't *Mu-ricella* and yet right there sitting on it was a minute pygmy sea-horse. Instantly, Denise knew it was something different. This wasn't a Bargibant's seahorse, she thought. It couldn't be. For a start it was sitting on the wrong type of sea fan. For the pygmies she already knew, it was *Muricella* or nothing. Then there was its size; this seahorse was a lot smaller than any other pygmy she had ever seen and much smoother and skinnier, without the portly belly she was used to seeing. It was also the wrong color, more orangey-yellow than lurid pink, and what's more, it was behaving very strangely for a Bargibant's. Instead of wrapping its tail around a branch of sea fan and sitting quiet and still, this little guy was fidgeting about, swimming up and down, around and around its sea fan home. The seahorse was not alone, but shared its sea fan with a gang of tiny transparent shrimp. Whenever the hyperactive seahorse wandered too close to a shrimp, it received a sharp nip on the head.

Following that dive, Denise started scrutinizing different species of sea fan and to her amazement found many more of these miniscule pygmies. They lived on various sea fans, often ones encrusted in other creatures like feather stars and brittle stars. She talked about them to anyone who would listen, utterly convinced that they were not just the offspring of the Bargibant's seahorses, which by then she had come to know so well.

It was following the publication of a feature article in the UK's *BBC Wildlife Magazine*, showcasing Denise's portraits of these odd-looking pygmies, that Amanda Vincent from Project Seahorse in Canada contacted her. Vincent was keen to find out more about these strange seahorses and gave Denise instructions on what to look for in order to narrow down whether this was indeed a new species. Were they all really that small? Did she ever see pairs dancing and courting together? Had she ever seen a pregnant male? After countless more dives, Denise answered yes to all these queries. She even caught sight of a male giving birth, sealing the deal that these were not immature juveniles of the slightly bigger Bargibant's seahorses but had to be a newly discovered species.

Nevertheless the scientists still wouldn't take her word for it; preserved specimens were needed. On a rare trip back home to America, Denise tried taking a couple of her unusual pygmies with her, but for the first time in all her travels, the airline lost her luggage, seahorses and all. She returned to Sulawesi and collected another set of specimens that finally found their way to John Randall, a fish taxonomist from the Bishop Museum in Hawaii. "If anyone would know what they were, he would," Denise said. And while Sara Lourie from Project Seahorse was working on her guide to all the seahorses of the world, she paid Denise a visit in Indonesia and saw the tiny pygmies in the wild.

Lourie and Randall, who had never met, collaborated over e-mail and phone to decide between them whether Denise was right that this was indeed a new species. These were the tiniest, fiddliest specimens either of them had worked with. They used special digital calipers under the gaze of a microscope to measure the seahorses to within a hundredth of a millimeter, less than half the width of an average human hair. The diagnostic bony ridges on their bodies and the rings of their tails were so soft and faint that the minute seahorses had to be put in an X-ray machine to see what was going on inside. But eventually, the seahorse sleuthing pair were agreed and in 2003 they published a paper describing a new species of seahorse. It was the smallest seahorse that had been found, reaching a mighty twenty-one millimeters when stretched out, about the diameter of a penny. Lourie and Randall decided to name it *Hippocampus denise*, Denise's pygmy seahorse, in honor of the lady who knew she was looking at something different.[26]

Since Denise's discovery, scuba divers around the world have been on the lookout for minute seahorses, and six more pygmies have so far been found. Some of them are even smaller than *Hippocampus denise* and all but one were named after the divers who spotted them. Coleman's pygmy seahorse, originally photographed by Neville Coleman, has so far only been found in

seagrass beds surrounding Lord Howe Island, six hundred kilometers offshore from the New South Wales coast in Australia.
Underwater photographer Helmut Debelius was instrumental
in tracking down the softcoral seahorse (*Hippocampus debelius*)
on the colorful reefs of the Red Sea. And it is thanks to the keen
eyes and patience of three scuba divers in Indonesia that a trio
of pygmies—Pontoh's, Severns' and Satomi's—have recently
come to light.[27]

If during a dive or a snorkel you are lucky and careful enough to
spot a seahorse and want to know what kind it is, the best thing
to do is take a few photographs. It has never been cheaper or
easier to have a go at being Jacques Cousteau or Jean Painlevé
using a compact digital camera snapped snugly inside a handy
plastic case. Failing that, you should try and remember whether
the seahorse you see is smooth or covered in spines, notably fat
or slim, and whether it has what you might decide is a stubby or
slender nose. Then make your best guess at how big it would be
from the top of its head to the tip of its uncurled tail, not forget-
ting that light bends as it passes into your diving mask, making
things appear to be a quarter bigger than they really are. Tempting as it may be, try not to focus too much on what color it is. A
"yellowy-orange seahorse" could be almost any species that was
feeling in a yellowy-orange mood the day you spotted it. But
do, however, look for any obvious stripes, bars, or saddles, since
these features can help distinguish one species from another.

Make a note of where you saw your seahorse, what sort of
habitat it was living in, and how deep, since that can narrow
down your list of possible species. Minotaur seahorses, for example, have only ever been seen in trawling nets hauled up from
the deep waters around Tasmania, not somewhere you are likely
to reach while diving. Other species, like Barbour's seahorses

and sea ponies, live in shallow seagrass meadows and mangrove-fringed lagoons, sometimes around man-made structures like jetties and marinas, rarely venturing deeper than ten meters.

Then, when you return to land, glance at a map of where you are, since geography will also let you cross off definite noes—not all seahorses live everywhere. There are only two species that live around the coasts of Great Britain, and across the Caribbean Sea there are three. However, if you are diving on riotously diverse coral reefs like those in Indonesia, Papua New Guinea, or Australia, your shopping list of possible seahorses is huge.[28]

Now, ready with your list of contenders, browse through photographs, drawings, and descriptions of seahorses in an up-to-date identification guide or Web site and, taking a deep breath, make a good guess at which one you think best fits the seahorse you met. And if, like Denise Tackett, Neville Coleman, Mike Severns, Hence Pontoh, or Satomi Onishi, you're convinced that none of them match, perhaps it's time you got in touch with the experts.

Our treasured journeys to visit the seahorses not only bring us closer to these addictive creatures but they have also opened our eyes to how their submerged world has been changing. If we could look down from above onto a stretch of ocean and hit the rewind button and scroll back through time, it would appear that nothing has changed; the rhythmic daily up-and-down heaving of the tides becomes a blur while the mantle of blue-green keeps hidden all that lies beneath. Scuba diving lets us tweak at the corner of that opaque blanket of ocean and peer underneath. There we are witnessing changes that matter not just for seahorses but for everything living in the ocean and the millions of people who depend on them.

Even if, as you read this sentence, the global fishing fleet were—somehow, for some reason—to catch its very last seahorse and then leave the rest of them alone forever, their future would still remain uncertain. The delicate habitats that seahorses rely on are becoming damaged, cut back, eroded, overheated, and depleted. Mangrove forests are felled to make way for marinas and shrimp farms; seagrass meadows are smothered in silt, pesticides, and fertilizers; the natural balance of coral reefs is being tipped upside down by too much fishing and pollution. The lurking metamorphosis of the oceans is felt strongest by the world's rarest seahorses.

Cape seahorses, also known as Knysna seahorses, do not look any different from most other members of the *Hippocampus* genus. As seahorses go, these are a stout variety with stubby snouts and they lack the weedy crown that would make them seahorse royalty. They are most often a mottled brown color but can, if they choose, change to yellow, green, beige, or black. What is unusual about Cape seahorses is that even though they are thought to be the world's rarest seahorses, they are not too difficult to find because there is only one place to look for them. Every single wild Cape seahorse lives in a patch of habitat that covers less than fifty square kilometers of sea, an area smaller than the island of Manhattan. Their numbers are uncertain, as is often the case with seahorses, but estimates indicate that there could be fewer than ninety thousand Cape seahorses living in the wild; that is one for every seventeen Manhattanites.

If you want to meet a Cape seahorse in the wild, drive eastward from Cape Town toward Port Elizabeth along South Africa's N2 route, also known as the Garden Route, and after several hours you will reach a small coastal town called Knysna. In the nineteenth century, Knysna was an important trading post for timber, ivory, and gold, but it is now best known as a popular spot for vacationers. Inland, hiking trails wander through hills shaded by cool indigenous forests of thousand-year-old yellowwood and stinkwood trees. Down at the shoreline, a pair of craggy sandstone cliffs guard shallow waters that lap absentmindedly on Knysna's idyllic sandy beaches. It is among the sandbanks, reed beds, and seagrass meadows that Cape seahorses seek shelter from the chaos of the Indian Ocean, in the relative safety of Knysna Lagoon.

With so few individuals living in such a small area—most of them in Knysna Lagoon and some in the nearby Swartvlei and Keurbooms estuaries—Cape seahorses have quite literally put all their eggs in one small basket. This makes their continued existence an inherently risky prospect. The arrival of a chance event like a severe storm or a hot dry summer could be all that is needed to wipe them all out. Indeed in 1991, following a sudden downpour, three thousand Cape seahorses were washed up dead on Knysna's beaches. The seahorses were asphyxiated when a breached sandbank sent estuary waters gushing into the Indian Ocean, leaving behind shallow water that quickly heated up in the African sun. Over three percent of the total population was lost in one go.[29]

But the real problem is not just the random events that Cape seahorses have to deal with. Their world is also being constantly chipped away a little piece at a time by the encroaching human world. People are also very much to blame for the Cape seahorse's World Conservation Union label of "Endangered."

Spiraling development of golf courses, resorts, houses and hotels, and the associated influx of people to an already over-crowded area, all are leaving an indelible footprint on the waters of Knysna Lagoon, where Cape seahorses are trying to keep a tail hold. Sediments and sewage pour in, seagrasses are mashed by sailing boats and water-skiers. In the Swartvlei estuary, when the water table rises above a certain threshold, the septic tanks of riverside homes stop working. Local authorities see the only remedy for this is to bulldoze the estuary mouth to let the water levels recede once more. When this happens, conservationists try to rescue the seahorses, transferring them to safer waters, but inevitably many are left behind.

There might not be much time to get it right if there are still going to be Cape seahorses in the wild in a few years' time. But there is at least some reason for hope. The residents of Knysna are starting to treasure the special feeling of having their very own seahorses living on their doorstep and nowhere else. Cape seahorses are now protected under South Africa's Sea Fisheries Act No. 58, which means they cannot be caught or disturbed. And since 1985, Knysna Lagoon has been designated as a multiple-use marine park, with special zones cordonned off for different activities, offering the seahorses at least some quieter areas to hide from the havoc of the seaside resort.

It is inside the invisible boundaries of marine parks, like the one at Knysna, that the greatest step forward in global marine conservation is taking place. Over the past twenty or thirty years, there has been mounting recognition that the most effective, not to mention the simplest, way of healing the oceans is merely to leave parts of them alone. All we need do is choose areas of sea where we declare no fishing, no extracting, no dumping, and no interference from humans of any sort. Although waters still flow through and fish will swim where they like, protecting fixed areas of seabed and the water column above it really does work. Creatures seek refuge, grow bigger and more abundant; habitats recover. Some fish and crustaceans will spill over into nonprotected areas and help boost catches elsewhere.

But what about the growing stockpile of carbon dioxide that

is building up in the atmosphere? Surely climate change does not adhere to the notional barriers of marine reserves or marine parks. The most catastrophic outcome of mankind's love affair with fossil fuels is likely to be a comprehensive, three-pronged attack on the oceans. As the insulating layer of carbon dioxide warms the planet's surface, the oceans, too, will heat up, causing sea levels to rise; if this happens too quickly, shallow marine ecosystems that depend on sunlight, including coral reefs and seagrass beds, might not be able to keep pace, and ultimately they will drown. Without reefs to protect them, shorelines and human coastal communities will become even more vulnerable to rising sea levels.

The second problem is that the ocean surface needs only to warm up by one or two degrees for many marine creatures to start getting dangerously hot and bothered. Mobile species have the advantage that they can simply move northward or southward into cooler waters, although their migrations are likely to trigger untold knock-on effects when they arrive uninvited in other ecosystems. The rising water temperature around the British Isles over the last forty years has caused species of zooplankton to march poleward by around ten degrees in latitude, the equivalent of twenty-five kilometers a year. This fundamental shift near the base of the oceanic food chain—since the majority of sea creatures eat either zooplankton or the predators of zooplankton—has sent ripples of change through the rest of the ecosystem. Sand eel populations have crashed, unable to follow their preferred zooplankton food as it moved north; consequently seabird populations, including puffins and guillemots, have collapsed because, without sand eels, they too are starving.[30]

Bottom-dwelling species that are rooted to the spot have no choice but to stay put as the water around them warms. When corals are bathed in seawater that is just slightly warmer than they are used to, the single-celled algae living inside them and providing most of their energy for some reason get irritated, pack their bags, and leave. The transparent corals are left naked and appear bleached and bone white, their calcium carbonate skeletons no longer covered by pigmented algae. Some corals cope for a while on their own and survive until another cohort of algae moves in and takes up the duties of food provision. Many corals are less resilient to coral bleaching and will die.[31] In the warming seas, the sea fans that Bargibant's and Denise's pygmy seahorses call home are simply melting away.[32]

As if that weren't enough, there is one more aquatic sting in the tail of climate change. Not only do higher temperatures spell disaster, but mounting levels of carbon dioxide also reap their own brand of turmoil. Imagine the bubbles of carbon dioxide that make a soft drink fizzy; some of the gas dissolves into the liquid, forming mild carbonic acid. The same thing happens in the oceans, only on a far more massive scale. Elevated levels of carbon dioxide in the atmosphere have already turned the oceans slightly more acidic than they were several decades ago. Perhaps this wouldn't be such a great disaster were it not for the fact that many creatures living in the ocean are essentially made of chalk. What happens when you put a piece of chalk into acid? It begins to dissolve away like an Alka-Seltzer in a glass of water. Laboratory tests have shown that corals, seashells, sea urchins, and—perhaps most worryingly—phytoplankton, which perform a crucial role of absorbing and locking away much of the world's carbon dioxide, all grow more slowly, and some not at all, when they are subjected to mildly more acidic conditions than "normal" seawater.

Take any one of these climate problems on its own and the results are bad enough, but with all three acting together, it comes as no surprise that prospects for the oceans are depressingly grim. Nevertheless, marine reserves can and do play their part. The effects of climate change may be global, but it does not rule out the importance of local and regional measures to help minimize them. Fish, for example, play a vital role in boosting coral reef resilience by munching their way through one of the reef's bad guys: seaweed. When corals are wiped out—by coral bleaching, say—large fleshy seaweeds often march in, taking up space on the reef and hindering coral recovery. An injection of nutrients from fertilizer runoff also tips the reef balance in favor of seaweeds, giving them extra strength to win the daily battle against corals. However, if shoals of herbivorous fish are protected inside marine reserves, they act as industrious gardeners, pruning unwanted seaweeds, keeping them in check and giving the corals a better chance of surviving when other problems come along. In general, ecosystems inside marine reserves will be healthier and much better equipped to deal with the stress of climate change, creating vital plots of seabed that are both more resistant to large-scale changes and more likely to bounce back.[33]

Over the last few decades, there has been a lot more talk about protecting the oceans. Back in 1992 at the Rio Earth Summit, hundreds of nations signed up to the Convention on Biological Diversity, which alongside many other conservation targets pledged to create a network of marine reserves covering at least ten percent of each different type of ecosystem found in the oceans.[34] The G8 nations talk about safeguarding even more of the marine realm, perhaps as much as twenty or thirty percent. But there is still a long way to go. Currently, less than one percent of the oceans are offered any sort of protection from the activities of mankind.[35]

Seahorses may be inconspicuous and small, they may hide in

quiet corners of the coast away from all but the keenest of eyes, but they can play an important role in encouraging us to protect parts of their vast ocean home. Increasingly, seahorses are being used as catalysts for conservation initiatives; they are being held aloft as poster species to help muster support for protecting the oceans. They are touchstones to remind people of the vulnerable, beautiful creatures that live there, giving us a reason to care. Not only that, but in many ways seahorses are the oceans' equivalent of miners' canaries. They are sensitive creatures, thriving only in undamaged habitats in unpolluted waters. Lose either of those things and seahorses will fade away, a warning perhaps of a wider catastrophe to come. They are the secret jewels in the crown of coral reefs and seagrass meadows, a barometer for the condition of the waters in which they live. The idea is that to save seahorses, you need to save the seas, or if you like, the other way around.

In 2005, a group of Malaysian marine biologists and journalists were outraged when they discovered that large areas of seagrass and mangroves were being uprooted along Peninsular Malaysia's east coast to make way for heavy industry, without so much as a murmur from local people. Initial surveys revealed that the Pulai River estuary in the state of Johor contained not only the country's largest seagrass bed but also Malaysia's only substantial population of yellow seahorses. Choosing the seahorses as a focus, a new organization was founded to help protect these important habitats: Save Our Seahorses (SOS Malaysia). So far, members of SOS have worked hard to lobby local government and policy makers, persuading them that seahorses and their habitats matter. Local fishermen used to collect seahorses in this part of Malaysia—in a similar way to the lantern fishermen of the Philippines—but already numbers are too low to make it profitable. Even though the people of Johor don't rely on seahorses for their income, SOS is using these captivating creatures to raise awareness and empower local communities, encouraging them to gain a sense of stewardship over their marine heritage and to play an active role in managing and protecting coastal habitats. In 2006, the organization rolled out a campaign to raise awareness of the seahorses' plight. The program brought more than two hundred Malaysian volunteers, many of them from cities, face-to-face with the seahorses and seagrasses of the Pulai estuary. Volunteers helped to conduct survey work, measuring seagrass habitat and tagging seahorses. And perhaps more important, they went back home with a newfound passion for the creatures living at the watery fringes of their country, with an impetus to pay attention to what is going on around them, to vote with their voices and start making a difference for the oceans.[36]

And just as disappearing seahorses can signify environmental gloom, so returning seahorses are an indication of ocean recovery. Improving conditions in one of England's busiest stretches of water has recently been heralded by the arrival of seahorses in London. Until recently, short-snouted seahorses were only very rarely spotted swimming up the Thames estuary.[37] One was seen in 2004, and before then, not since 1976. But in 2008, a healthy population of seahorses was found in the estuary, setting up home as far inland as Dagenham, in the eastern reaches of Greater London and still within reach of salty tides.[38] Historically, the Thames has been notoriously polluted, a dumping ground for the inhabitants of England's capital city. But cleanup efforts have been paying off, not only with seahorses but also with salmon, porpoises, seals, and otters all making a comeback. Some researchers think seahorses may have been enticed north from their usual stronghold around the Mediterranean because global warming has taken the chill off North Sea waters. But nevertheless, their presence is testament to the recovering health of Thames riverine habitats, and Londoners are both proud and intrigued by their new aquatic neighbors. In January 2006, a seven-ton northern bottlenose whale hit the headlines when it wandered into the Thames and got lost. For a few days the riverside in the heart of central London thronged with eager whale watchers, keen to catch their own glimpse of the exotic visitor. The city mourned when the whale—whom the British newspapers variously nick-named Willy, Whaley, and Wilma—died, sadly, during attempts to rescue her.[39] But unlike the tale of the doomed Thames whale, this seahorse story, so far, has a much happier ending.

EPILOGUE

$\mathcal{U}$ltimately though, does it really matter if there are seahorses or not? It almost certainly doesn't matter for the ecology of the oceans. Seahorses may rely on a healthy ocean to survive, but the oceans don't rely on them. They are not what ecologists would call "keystone" species, a term reserved for such things as giant fig trees in tropical rain forests, which feed a multitude of creatures with their leaves, fruit, pollen, and nectar. Take away the fig trees and you might also lose monkeys, gibbons, hornbills, toucans, bats, moths, and wasps.[1] Seahorses may occasionally show up in the stomachs of tuna and even penguins, but no one creature relies on them as part of a staple diet. Nor do they form the basis of complex habitats like reef-building corals or dense beds of oysters that create homes for thousands of other species. And their appetites may be greedy for creatures so small, but they don't perform an important role in controlling populations of other species. Some animals and plants are sorely missed when they disappear. In the nineteenth century, Alaskan sea otters were hunted for their fur to within inches of extinction, which triggered a wave of disruption resonating through the entire marine ecosystem. Without predatory otters to eat them, sea urchin populations boomed, unleashing spiny plagues so voracious, they stripped coastlines of their luxuriant kelp forests.[2] Nothing like that would happen if there were no more seahorses. Without seahorses, the oceans will still work just as well—or just as poorly—as they did with them.

Even if nature wouldn't pay much attention the day seahorses were no longer there, surely they do still matter to us. They matter to scientists because they offer insights into life beneath the waves and quietly answer questions about what it means to be male. They obviously matter to the people whose lives revolve tightly around them, the fishers, traders, aquarium keepers, and medicine makers. And they matter because they inspire us to care about the natural world.

But more than anything, if wild seahorses were to fade away, perhaps we would all lose something important in our own minds. Sir David Attenborough, Britain's—and probably the world's—most cherished and highly regarded natural history broadcaster, once said that the reason we should protect tigers or seahorses or any of the planet's biodiversity is not because there would be some sort of eco-disaster without them. Instead, as he said, "The overwhelming reason is man's imaginative health."[3]

There is something that remains intangible yet intoxicating about the idea of seahorses. For a living creature to look so strange and yet so perfectly pleasing at the same time, we might assume it must be a magnificent feat of deliberate design, a fairy tale made real. And yet what great satisfaction is to be had from knowing that the unlikely seahorses are merely a result of the unseen forces of natural selection at work.

I have seen several more wild seahorses since my first encounter with the orange one with white saddles in Vietnam, but since then I have realized two things. First, I now know that the possibility of a seahorse will forever tempt me and I will never stop looking for them. Whether I am snorkeling or diving or poking around a rock pool, there will always be a tiny part of me still wishing that a seahorse will be waiting for me behind the next clump of seaweed. And yet I've also realized that even if I never see a wild seahorse again, it wouldn't be so bad. To see one is to contemplate one, to pause briefly in blissful tranquility and wonder why and how. But that memory doesn't fade and the recollection of a single seahorse is enough to last a lifetime. In the end, all that really matters is that they are still out there, somewhere. I expect there are many people who spend their whole lives in or next to the sea and who will never be in the right place at the right time to see a seahorse. But the world is absolutely a better place just knowing there are seahorses swimming through the oceans.

Imagine what it would be like if all we had to tell our grandchildren were stories of a time when there used to be wonderful creatures called seahorses living wild in the oceans. They looked like miniature horses with rolling eyes and tiny monkeys' tails. It was the males that had babies—no animals do that anymore—and they changed color as if by magic and danced elegant dances every day with their faithful partners. If stories were all that were left of the seahorses, I don't suppose anyone would believe us.

ACKNOWLEDGMENTS

*I*n 1955, Rachel Carson began the acknowledgments of her book *The Edge of the Sea* with "Our understanding of the nature of the shore and of the lives of sea animals has been acquired through the labor of many hundreds of people, some of whom have devoted a lifetime to the study of a single group of animals." Nothing could be truer for the seahorses, and the rest of the Syngnathidae family, whose strange lives I have attempted to capture in the pages of this book. The names of many scientists appear in the bibliography and to each one I am greatly indebted for their own particular devotion to this peculiar group of species that I have traced through the academic literature. For their welcome advice, information, and guidance I especially wish to thank Heather Koldewey, Choo Chee Kuang, Neil Garrick-Maidment, Rudie Kuiter, David Harasti, Denise Tackett, Helmut Debelius, Douglas Herdson, and Brian Allanson. Thank you also to Elizabeth Call for her advice on the theories and practices of Traditional Chinese Medicine. And for their insights into the world of seahorse farming and keeping, I would like to thank Rachel Hawkins of Seahorse Australia, Tracy Warland of the South Australian Seahorse Marine Service, David Warland from syngnathid.org, Helen Gill from Simply

Seahorses, and the members of various online seahorse-keeping forums—in particular Carmine Scotch, Kells, and Deanne Ward. Many thanks to Micki Scales for her help with translations. At TRAFFIC East Asia I want to thank Craig Kirkpatrick, Caroline Liou, and Joyce Wu. I would also like to thank the staff at the University Library and the Central Science Library in Cambridge, and the folks at the British Library in London.

I am deeply grateful to all the people who made my overseas research trips both possible and fun. In Mauritius, I want to thank Tony Apollon and all the dive masters at the Coral Diving Centre for kindly taking me diving and joining in with the seahorse hunt. Thanks also to Olivier Tyack and members of the Mauritius Underwater Group and the Mauritius Marine Conservation Society for their warm welcome back in 2005. In Vietnam my thanks go to Truong Si Ky, Do Huu Hoang, and Bui Thi Minh Ha at the Institute of Oceanography in Nha Trang for their hospitality and for making the impossible possible by arranging for my trip overnight onboard a trawling boat on the South China Sea. And a great thank you to Hoang Ngnyen from Rainbow Divers in Nha Trang for helping me find my first wild seahorse. In Belize, I want to say a very warm thanks to Vicki Snaddon for letting me move into her glorious beachside penthouse, where much of chapter 6 was written. And thank you also to all the people of Ambergris Cay and Lighthouse Reef who kept their eyes peeled for seahorses for me. I am hugely grateful to Kathlina Alford in Chattanooga for showing me around the Tennessee Aquarium and to Todd Stailey for letting me use his photographs from the aquarium. Thank you to my sister Ruth for being great company on my trip to Scotland, and in Orkney, I want to thank Alan Jackson and Cindy De Battista at the Orkney Marine-Life Centre, Anne Brundle at the Tankerness House Museum, and Alistair Skeene and Jon Side.

And although I sadly didn't make it there to visit the seahorses, I am grateful for Lyn Costenaro at Sea Saba for her insights into Caribbean seahorse diving.

This book would have stayed firmly inside my head were it not for a group of people who thought that my ideas were worth writing down, and who never let me give up. I am indebted to Ben Hesketh for saying the right things at the right time. I am truly indebted to my agents Eva Talmadge and Emma Sweeney for their perfect combination of kind support, endless patience, and constructive advice and simply for being so great to work with. I must also say a huge thank you to all the people at Gotham Books and especially my editor Jessica Sindler for all her enthusiasm, belief, and encouragement all the way through the writing and production of this book. I also want to wish a huge thank you to all my family: to my parents, Tom and Diana Hendry, and to my sisters, Ruth and Kate Hendry, who all read many of my words as I went along. And thank you to the Scales tribe for putting up with a lot of talk about seahorses. Thank you to Sam and David Thorp for lending me their knowledge of cartoons and to my friends who willingly and graciously took on the role of readers: Ria Cooke, Anna Petherick, Tom Fayle, Katie Boswell, John Osborn, and Cerian Neal. And to my husband, Ivan, we all know this would never have happened without you, if you hadn't been there to look after me, to help me search for seahorses, to pour me a hot bath when things got tough, and to give me all the gentle persuasion I needed to carry on. Thank you for not insisting that I go out and get a proper job.

Appendix

MEET THE SEAHORSES

Below are the main syngnathids that appear in the book, including common name, Latin name, name and date of original discovery, maximum height, and main geographical range.

Seahorse species named in Lourie et al 2004

Big-belly seahorse, pot-bellied seahorse. *Hippocampus abdominalis* (Lesson 1827). 35cm. Australia, New Zealand.

West African seahorse. *Hippocampus algiricus* (Kaup 1856). 19cm. Angola to Sierra Leone.

Narrow-bellied seahorse, western spiny seahorse. *Hippocampus angustus* (Günther 1870). 16cm. Australia.

Barbour's seahorse, zebra-snout seahorse. *Hippocampus barbouri* (Jordan and Richardson 1908). 15cm. Indonesia, Malaysia, Philippines.

Bargibant's seahorse, pygmy seahorse. *Hippocampus bargibanti* (Whitley 1970). 2.4cm. Australia, New Caledonia, Indonesia, Japan, Papua New Guinea, Philippines.

Réunion seahorse. *Hippocampus borboniensis* (Duméril 1870). 14cm. Réunion, Madagscar, Mauritius, East African coast between Tanzania and South Africa, possibly the Comoros Islands.

Knobby seahorse, short-snouted seahorse, short-headed seahorse. *Hippocampus breviceps* (Peters 1869). 10cm. Australia.

Giraffe seahorse. *Hippocampus camelopardalis* (Bianconi 1854). 10cm. Mozambique, Tanzania, South Africa.

Cape seahorse, Knysna seahorse. *Hippocampus capensis* (Boulenger 1900). 12cm. South Africa.

Tiger tail seahorse. *Hippocampus comes* (Cantor 1850). 18.7cm. Southeast Asia.

Crowned seahorse, horned seahorse. *Hippocampus coronatus* (Temminck and Schlegel 1850). 12.7cm. Japan.

Denise's pygmy seahorse. *Hippocampus denise* (Randall and Lourie 2003). 2.14cm. Southeast Asia, Micronesia, Palau, Papua New Guinea, Solomon Islands, Vanuatu.

Lined seahorse, northern seahorse. *Hippocampus erectus* (Perry 1810). 19cm. Eastern seaboard of North America from Newfoundland south, throughout the Caribbean, to the eastern seaboard of South America, possibly as far as Uruguay.

Fisher's seahorse. *Hippocampus fisheri* (Jordan and Evermann 1903). 8cm. Hawaii.

Sea pony, drab seahorse. *Hippocampus fuscus* (Rüppell 1838). 14.4cm. Djibouti, India, Saudi Arabia, Sri Lanka.

Long-snouted seahorse, hairy seahorse. *Hippocampus guttulatus* (Cuvier 1829). 18cm. Mediterranean Sea, Atlantic coast of Europe from Portugal and Spain to the southern shores of the UK.

Short-snouted seahorse, Mediterranean seahorse. *Hippocampus hippocampus* (Linnaeus 1758). 15cm. Mediterranean Sea,

Atlantic coast of Europe from Portugal and Spain to the northern shores of the UK.

Thorny seahorse. *Hippocampus histrix* (Kaup 1856). 17cm. India, China, Japan, Southeast Asia, Papua New Guinea, Micronesia, Tonga, Hawaii, Mauritius, Tanzania, South Africa.

Pacific seahorse. *Hippocampus ingens* (Girard 1859). 31cm. Western seaboard of the United States from California south through Central America to Peru.

Jayakar's seahorse, spiny seahorse. *Hippocampus jayakari* (Boulenger 1900). 14cm. Israel, Oman, Pakistan.

Kellogg's seahorse, great seahorse, offshore seahorse. *Hippocampus kelloggi* (Jordan and Snyder 1902). 28cm. Pakistan, India, China, Japan, Southeast Asia, India, Tanzania, Australia.

Yellow seahorse, spotted seahorse, estuary seahorse. *Hippocampus kuda* (Bleeker 1852). 17cm. Pakistan, India, China, Japan, Southeast Asia, Australia, Papua New Guinea, Micronesia, New Caledonia, Solomon Islands, Tonga, Hawaii.

Lichtenstein's seahorse. *Hippocampus lichtensteinii* (Kaup 1856). 4cm. Red Sea.

Bullneck seahorse, Minotaur seahorse. *Hippocampus minotaur* (Gomon 1997). <5cm. Australia.

Lemur-tail seahorse, Japanese seahorse. *Hippocampus mohnikei* (Bleeker 1854). 8cm. Japan.

Slender seahorse, long-snout seahorse. *Hippocampus reidi* (Ginsberg 1933). 17.5cm. Caribbean from Bahamas south to Venezuela.

Shiho's seahorse. *Hippocampus sindonis* (Jordan and Snyder 1902). 8cm. Japan.

Hedgehog seahorse. *Hippocampus spinosissimus* (Weber 1913). 17.2cm. Southeast Asia to Australia.

West Australian seahorse, tigersnout seahorse. *Hippocampus subelongatus* (Castlenau 1873). 20cm. Southwest coast of Western Australia.

Three-spot seahorse, low-crowned seahorse, flat-faced seahorse. *Hippocampus trimaculatus* (Leach 1814). 17cm. India, China, Japan, Southeast Asia to Australia.

White's seahorse, New Holland seahorse, Sydney seahorse. *Hippocampus whitei* (Bleeker 1855). 13cm. Southeast coast of Australia.

Zebra seahorse. *Hippocampus zebra* (Whitley 1964). 9.4cm. Australia.

Dwarf seahorse. *Hippocampus zosterae* (Jordan and Gilbert 1882). 2.5cm. Bahamas, Mexico, southern United States.

Additional species named in Kuiter 2001

Winged seahorse. *Hippocampus alatus* (Kuiter 2001). 13.6cm. Northwest coast of Australia.

False-eye seahorse. *Hippocampus biocellatus* (Kuiter 2001). 6cm. Shark Bay, Australia.

Southern pot-belly seahorse. *Hippocampus bleekeri* (Fowler 1908). 23cm. Southern Australia.

Low-crown seahorse. *Hippocampus dahli* (Ogilby 1908). 11.6cm. Northeastern Australia.

West Australian seahorse. *Hippocampus elongatus* (Castlenau 1873). 14.5cm. Western Australia.

Big-head seahorse. *Hippocampus grandiceps* (Kuiter 2001). 5.8cm. Northern Australia.

Eastern spiny seahorse. *Hippocampus hendriki* (Kuiter 2001). 8.7cm. Northeastern Australia.

Collared seahorse. *Hippocampus jugumus* (Kuiter 2001). 4.4cm. A single specimen collected from Lord Howe Island, Australia.

Smooth seahorse. *Hippocampus kampylotrachelos* (Bleeker 1854). 22cm. Sumatra, Bali, and Timor Sea.

Monte Bello seahorse. *Hippocampus montebelloensis* (Kuiter 2001). 4.1cm. Western Australia.

Northern spiny seahorse. *Hippocampus multispinus* (Kuiter 2001). 8.6cm. Northern Australia.

Flat-face seahorse. *Hippocampus planifrons* (Peters 1877). 5.2cm. Shark Bay and Broome, Australia.

High-crown seahorse. *Hippocampus procerus* (Kuiter 2001). 7.9cm. Queensland, Australia.

Queensland seahorse. *Hippocampus queenslandicus* (Horne 2001). 11.6cm. Queensland, Australia.

Half-spined seahorse. *Hippocampus semispinosus* (Kuiter 2001). 8.9cm. Specimens caught in trawls in southern Indonesia and northwestern Australia.

Common seahorse. *Hippocampus taeniopterus* (Bleeker 1852). 20cm. Indonesia, Papua New Guinea, northeastern Australia.

Sad seahorse. *Hippocampus tristis* (Castlenau 1872). 22.2cm. Queensland, Australia.

Knobby seahorse. *Hippocampus tuberculatus* (Castlenau 1875). 4cm. Western Australia.

Seahorses named subsequently to Lourie et al 2004

Coleman's pygmy seahorse. *Hippocampus colemani* (Kuiter 2003). 2.3cm. Lord Howe Island, Australia.

Pontohi's pygmy seahorse. *Hippocampus pontohi* (Lourie and Kuiter 2008). 1.4cm. Indonesia.

Satomi's pygmy seahorse. *Hippocampus satomiae* (Lourie and Kuiter 2008). 1.3cm. Indonesia, Malaysia.

Severns' pygmy seahorse. *Hippocampus severnsi* (Lourie and Kuiter 2008). 1.3cm. Indonesia, Papua New Guinea, Solomon Islands, Fiji, Japan.

Softcoral seahorse. *Hippocampus debelius.* (Gomon and Kuiter 2009). 1.7 cm. Red Sea.

Walea pygmy seahorse. *Hippocampus waleananus.* (Gomon and Kuiter 2009). 1.8cm. Walea Island, Indonesia.

Other syngnathids and close relatives *mentioned in* Poseidon's Steed

Yellow-scribbled pipefish. *Corythoichthys intestinalis* (Ramsay 1881). 16cm. Borneo to Samoa, Marshall Islands, North West Australia, New Caledonia.

Banded pipefish. *Doryrhamphus dactyliophorus* (Bleeker 1853). 19cm. Red Sea to East Africa, Japan to Australia.

Bluestripe pipefish. *Doryrhamphus excisus* (Kaup 1856). 7cm. Across Indian and Pacific oceans.

Flagtail pipefish, Negros pipefish. *Doryrhamphus negrosensis negrosensis* (Herre 1934). 4.7cm. Borneo to Vanuatu, Australia to Micronesia.

Ribboned pipefish. *Haliichthys taeniophorus* (Gray 1859). 30cm. Indonesia, North West Australia.

Worm pipefish. *Nerophis lumbriciformis* (Jenyns 1835). 15cm. British Isles, northern France to Scandanavia.

Leafy seadragon. *Phycodurus eques* (Günther 1865). 35cm. Southern Australia.

Weedy seadragon. *Phyllopteryx taeniolatus* (Lacepède 1804). 46cm. Southern Australia.

Ornate ghost pipefish. *Solenostomus paradoxus* (Pallas 1770). 12cm. Red Sea and East Africa to Fiji, Japan, Australia, Tonga.

Alligator pipefish. *Syngnathoides biaculeatus* (Bloch 1785). 29cm. Red Sea to South Africa, Japan to Australia.

Chocolate pipefish. *Syngnathus euchrous* (Fritzshe 1980). 25cm. Eastern Pacific, California to Mexico.

Bend-stick pipefish. *Trachyrhamphus bicoarctatus* (Bleeker 1857). 40cm. Red Sea to East Africa, Japan to New Caledonia and Micronesia.

NOTES

Prelude

1 For any readers who are familiar with British diving, yes, my first open-water dive was indeed at the National Dive Centre at Stoney Cove in Leicestershire. It was in March and the water was four degrees centigrade (thirty-nine degrees Fahrenheit). I was wearing a seven-millimeter semi-dry suit and on my first dive I lasted twenty minutes, the second one only fifteen. For those of you who would like to sample the joys of British inland diving, see www.stoneycove.com.

2 Years after finishing my Ph.D., I heard the depressing news that the herd of adult Napoleon wrasse I spent months swimming with and individually identifying at Layang Layang had all been carted off to the live fish market. Despite unofficial protection granted by the presence of the Malaysian military—keen to stake the nation's claim on this part of the disputed Spratly Islands—fishermen had been allowed in and helped themselves, probably catching the fish while they spawned. The only remaining substantial populations of Napoleon wrasse in the world may now be on the east coast of Malaysian Borneo, in the Maldives (where they are fully

protected), and in Palau in the central west Pacific. See www
.humpheadwrasse.info.

3 My year working in Malaysia was with the World Wide Fund
for Nature (WWF), when I surveyed the Langkawi archipelago, off the western state of Kedah.

4 Coral bleaching in Belize: Handwick and Hafvenstein
2003.

5 The ornate ghost pipefish (*Solenostomus paradoxus*) is a
member of the Solenostomidae family, a sister group to the
Syngnathidae: Kuiter 2009.

6 "so like the picture of a Saint George his dragon": Eastman
1915.

7 Development of hippocampi in the brains of London taxi
drivers: Maguire 2000. For more information on the knowledge test for London cabdrivers, see www.taxiknowledge
.co.uk.

Chapter 1

1 Discovery of the Uşak tomb: Norman 1993.

2 Source of Lydia's luxurious wealth: Lloyd 1989.

3 Cornell University archaeologist Andrew Ramage was a
graduate student at Harvard University in 1968 when he first
discovered the remains of a gold refinery in Sardis. A Cornell-
Harvard expedition, in collaboration with the British Museum, took many years to fully excavate the site. Ramage and
colleagues discovered that the Lydians had been the first
people to develop a process for rendering alluvial gold from
the Pactolus River into its component parts of silver and gold:
Ramage and Craddock 2000.

4 The "Lydian Hoard" is in fact an assemblage of artifacts from
several tombs in the Uşak region of Turkey. Some archaeolo-

gists think the hoard had nothing to do with King Croesus but was associated with the burial site of a high-ranking Lydian female: www.about-turkey.com/karun/toptepe.htm.

5 Dreamtime and Rainbow Serpent spirit ancestors: Isaacs 2005.

6 Theories of similarities between Rainbow Serpents and seahorses: Tacon, Wilson and Chippindale 1996.

7 Appearance of hippocampus in early European art, including theories of their Asian origins: Shepard 1940.

8 Single appearance of Hippocampus in Egyptian art: Murray 1911.

9 Background on Greek mythology: Hard 2004.

10 Adoption of Greek gods by the Romans: Morford and Lenardon 2007.

11 Richard Hobbs' investigations of Dahl's portrayal of the Mildenhall Treasure story and other theories: Hobbs 1997.

12 Background on Pictish stones: Mack 1997; Henderson and Henderson 2004.

13 Theories of the origins of Pictish Beasts from discussions with Anne Brundle from the Tankerness House Museum, Kirkwall, and from Cessford 2005.

14 Philip Pulmann's mulefa: Pulmann 2001.

15 Discovery of Tankerness seahorse: Mack 1997.

16 Details of the Pictish stone carving at Aberlemno: Lavergne and Ashmore 2001.

17 Dog-headed fish monsters: Allen and Anderson 1903.

18 Theory of hippocampus evolving into Pictish style seahorses: Cessford 2005.

19 Pictish animal carvings as lineage totems: Jackson 1993.

20 Seahorses as symbols of guardianship representing a protective role: Henderson 1997.

21   Distribution of seahorses in northern Scotland: Garrick-Maidment 2007.

22   For information on recent sightings of seahorses in Orkney (and lack of them), I interviewed Alan Jackson from the Orkney Marine-Life Aquarium and other local residents.

23   European bestiaries and theories of land-sea counterparts: White 1984.

24   A digitized version of the Aberdeen Bestiary, along with a transcription and translation, is available online at www .abdn.ac.uk/bestiary.

25   The long-held belief that male lions brought their dead newborn cubs to life may be a misinterpretation of the common practice of infanticide. When an outside male invades a pride, he will often kill any cubs of the defeated male, triggering the lionesses to return to heat and allowing the new male to breed immediately.

26   "it may very possibly be true": Pliny the Elder, *The Natural History* (Bostock and Riley trans.) 1842.

27   Legends of Manannán mac Lir: Howel 2002.

28   Stories of Kelpies and the Each Uisge: Briggs 1977.

29   Details of the lead up to and progression of the first Lydian Hoard court case between the Metropolitan Museum and the Turkish government: Rose and Acar 1995.

30   "We took our chances when we bought the material": Hoving 1993.

31   Acar, Kalyan, and Hoving's involvement in the Lydian Hoard case: Norman 1991.

32   Details of 2006 Turkish court case: Anon 2006, "Ten charged in missing brooch case"; Arsu 2006.

33   "You can secure the outside of the museum as much as you want": Erdem 2006.

Chapter 2

1 Mexican Seri legend: Smith 1958.

2 For an overview of the *Hippocampus* genus including habitat preferences, see Lourie et al 2004. Seahorses living on swimming nets: Clynick 2008. For an overview of the Syngnathidae family, see Kuiter 2009 (N.B. Rudie Kuiter includes seahorse species that others consider to be synonyms); www.fishbase.org/Summary/FamilySummary.cfm?ID=248.

3 The term *zooplankton* refers to the marine or freshwater animals that drift through the water column, and it stems from Greek words *zoon* meaning "animal" and *planktos* meaning "drifter" or "wanderer." Most are tiny and cannot be seen with the naked eye, but some are large, such as jellyfish. *Phytoplankton* (from Greek word *phyton* meaning "plant") are the members of the plankton that generate their own energy by photosynthesis and are mostly single-celled algae and blue-green algae.

4 For a summary of seahorse life history characteristics, see Foster and Vincent 2004.

5 Pipefish are the most diverse group within the Syngnathidae family. There are thought to be forty-two genera and close to two hundred species of pipefish that live in marine, brackish, and freshwaters. There are four genera and eleven species of pipehorse. The families closely related to Syngnathidae are trumpetfish (Aulostomidae), flutemouths (Fistulariidae), tubesnouts (Aulorhynchidae), seamoths (Pegasidae), sticklebacks (Gasterosteidae), bellowsfishes (Macroramphosidae) and ghost pipefish (Solenostomidae). The orange hairy ghost pipefish has not yet been officially named: www.austmus.gov.au/fishes/students/focus/soleno.htm.

6 Early appearance of seahorses in print: Matthioli 1565.

7   Rapid movement of seahorse fins: Ashley-Ross 2002. The frequency at which seahorses beat their fins is higher than the "flicker fusion threshold" in humans, making it seem that seahorses are drifting unpowered through the water.

8   "these slow pokes": Cousteau 1979.

9   A closely related family to the Syngnathidae are the shrimp-fish or Centriscidae, which, like seahorses, also swim upright but in a heads-down rather than heads-up position. They hang together in small shoals that mimic blades of seagrass or the spines of sea urchins; see Kuiter 2009.

10   "a sort of evolutionary harking back": Carson 1955.

11   Chameleons in West African folklore, personal communication from Alexandra Boswell.

12   Seahorse and pipefish feeding mechanisms: Lussanet and Muller 2007. Pipette feeding only works for predators that target small prey. Larger prey would require larger mouths to fit into, leaving the fish with a chunky snout that is too cumbersome for the speedy movements required for pipette feeding. Seahorses have been seen tackling larger prey by biting them into smaller pieces before swallowing.

13   "unexpectedly a sharp little noise" and Dufossé's study of seahorse communication: Gill 1905. Recent study of seahorse sound production: Colson et al 1998. This study also involved the surgical removal of the bony ridge at the rear of the skull (the supraoccipital ridge), resulting in a decreased production of clicks when the seahorse was feeding, supporting theories that seahorse vocalizations are stridulatory in origin and produced by the supraoccipital-coronet articulation.

14   Description of seahorse courtship and mating, evolution of male pregnancy and pouch morphology: Vincent 1990. Some pipefish are described as undergoing male pregnancy, but their broad pouches are formed by two ventral flaps glued together.

15 Descriptions of seahorse mating and birth: Whitley and Allan 1958.

16 Paternal care by the marsupial toad: Attenborough 2007, episode 2, *Land Invaders—Amphibians.*

17 Details of seahorse "aquasperm" and sperm duct being located outside the pouch: Van Look et al 2007.

18 *Linnaeus* is the Latinized form of the name Carl von Linné, by which he was known after his ennoblement.

19 Linnaeus' system of taxonomy works so well because it not only provides a neat way of naming every living thing but with each group nestled within a larger one (species within genus, genus within order), it gives an intuitive way of indicating how each creature is related to all other living things, arranged, as they are, along the bifurcating branches of an enormous tree of life.

20 "Linnaeus' disciples": Pratt 1992.

21 The Swedish East India Company (Svenska Ostindiska Companiet or SOIC) was established in Gothenburg in 1731 by Scottish trader Colin Campbell and was inspired by the success of similar organizations, the Dutch East India Company and the British East India Company.

22 Located on the island of Java, Batavia, established in 1611, was the second major trading post set up by the Dutch East India Company, whose territories later became the Dutch East Indies until the Republic of Indonesia declared independence in 1945.

23 Story of Bleeker in Indonesia: Nowak 2007.

24 Of the seahorses Bleeker originally identified, three are included in Lourie et al's list of thirty-three species: yellow seahorse, White's seahorse, and the lemur-tail seahorse. Australian seahorse expert Rudie Kuiter asserts that Bleeker was also correct in identifying the smooth seahorse and the common seahorse: Kuiter 2001.

25 Discovery of a bright orange seahorse off the southern coast of England: Anon 2006, "Island seahorse drifts off course."

26 For the recent update of seahorse taxonomy, see Lourie et al 2004. The decision to conduct a global revision of seahorse species was driven in part by the 2004 listing of the seahorse genus on CITES (see chapter 4), since identification of species in trade is a stipulation of the convention.

27 Description of six seahorses following Lourie et al 2003: Kuiter 2003; Lourie and Kuiter 2008; Gomon and Kuiter 2009.

28 Other seahorse genera have been previously proposed, including *Farlapiscis*, named after a famous racehorse Phar Lap, *Hippohystrix*, and *Macleayina*: Whitley and Allan 1958.

29 The term *ecology* was coined by German biologist, philosopher, and physician Ernst Haeckel in 1866, but the science of ecology didn't really get going until 1895 when Danish botanist Eugenius Warming published the first textbook on plant ecology, O*ecology of Plants*. http://en.wikipedia.org/wiki/Eugenius_Warming.

30 Cytochrome b is a molecule containing iron that is found in mitochondria, the energy-producing units that lurk inside living cells. It forms part of the electron transfer chain in the process of respiration during which cellular energy is generated from glucose. http://en.wikipedia.org/wiki/Cytochrome_b.

31 Molecular study of Vietnamese seahorses: Lourie et al 1999.

32 Second study of seahorse genetics using cytochrome b, showing problems with some species identification: Casey et al 2004.

33 Discovery of fossil seahorses in Slovenia: www2.arnes.si/~jzaloh/fossil_seahorses.htm.

34 The age of Italian seahorse fossils and study identifying

Bargibant's seahorses as the oldest species: Teske, Cherry and Matthee 2004.

35 In 2001, Rudie Kuiter published a revision of Australian sea-horses detailing many species that Sara Lourie and colleagues did not include in the 2004 identification guide, and nine completely new species. At a meeting of the CITES Nomen-clature Committee in 2003, discussions were held on the standardization of seahorse taxonomy in which it was recog-nized that many more species remain to be properly identi-fied but pointed out inconsistencies and limitations in some of Kuiter's methodologies, such as the description of some new species based entirely on a single specimen: CITES NC3 Doc 5 (18 August 2003). www.cites.org/common/com/ NC/fauna/NC3-05.pdf.

36 The discovery of fossil seahorses in an Australian vineyard: www.wakefieldwines.com.au/aboutUs.php.

37 Lack of evidence of seahorses beginning in Caribbean or Indo-Pacific, and the rough estimate of seahorses evolving 16.5 million years ago: Casey et al 2004. For some reason, this second study did not include Bargibant's seahorses, but both studies do at least agree on the great age of the other two species, the knobby and big-belly seahorses.

38 It seems odd perhaps that ancient seahorses have apparently left so few fossils behind. We might expect their external body armor would have assisted in the fossilization process, preserving the original shape of the dead seahorse and pre-venting rapid deterioration of the body.

39 Support for seahorse rafting hypothesis: Teske et al 2005. Other marine species are known to raft on clumps of sea-weed: Holmquist 1994.

40 West African seahorses probably rafted all the way from the Caribbean where their closest genetic relative, the slender

seahorse, lives. Short-snouted seahorses may also have waved good-bye to the Caribbean and sailed east across the Atlantic, when the Gulf Stream was stronger around four million years ago, to reach their current home in Europe: Teske et al 2007.

41 Evidence for ancient exploitation of giant clams in the Red Sea: Richter et al 2008.

42 Ninety-thousand-year-old remains of fishing gear: Yellen et al 1995.

Chapter 3

1 Legend of Shenonngjia mountain: Laidler and Laidler 1996.

2 Background on legends of Shen Nong and early development of Chinese herbal medicines: Needham 1984; www .shen-nong.com.

3 Modern reconstruction of the *Shen Nong Ben Cao Jing*: Yang 1998.

4 It was common practice for Chinese writers to include lengthy passages and even whole chapters of other people's works in their own, indicating quotes in different colored ink.

5 First appearance of seahorse in the *Ben Cao Gang Mu*: Vincent 1996.

6 Details of the continuing medical and spiritual uses of seahorses and other syngnathids: Vincent 1996; Baum and Vincent 2005; Marichamy et al 1993; Vincent 1997; Bensky and Gamble 1986; Martin-Smith and Vincent 2006; Martin-Smith, Lam and Lee 2003; Choo and Liew 2005; Smith 1958. Some say the reason some people believe seahorses can assist during childbirth and bestow good luck on children is because their curled shape resembles a fetus.

7 Seamoths, also known as sea sparrows, belong to the order Pegasidae, a close relative of the Syngnathidae family. There are two genera of seamoth, *Pegasus*, named by Linnaeus, and *Eurypegasus*, named by Bleeker: Kuiter 2000.

8 Details of Vietnamese fishermen's uses of seahorses were collected during a visit to Nha Trang, Vietnam.

9 Story of Zanzibar fishers using seahorses to bestow good fortune on their fishing nets: McPherson and Vincent 2004.

10 Seahorses as a horse tonic in Indonesia: Bennett 1834.

11 Account of P'ing Yi's water-fairy pills: Read 1939.

12 The oldest surviving copy of Dioscorides' *Materia Medica:* http://en.wikipedia.org/wiki/Vienna_Dioscorides.

13 Background on Dioscorides: Griggs 1981; Riddle 1985.

14 Lichen planus is a localized skin condition that causes a recurrent itchy rash. The precise cause is not known. The "seacalf" refers to a seal, "mæna" is a type of fish, "aoboli" is an ancient measure of weight, and "pastinaca" is a stingray.

15 Background on Aelianus: http://en.wikipedia.org/wiki/Claudius_Aelianus.

16 "Washed out the stomachs of the sea-horses, some of which he roasted" and "for the man who has tasted it is first of all seized with the most violent retching": Aelian, *On the Nature of Animals*, Scholfield trans. 1958.

17 Ciguatera poisoning occurs when people eat fish contaminated with a naturally occurring poison known as ciguatoxin, which originates in single-celled algae and accumulates in predatory fish near the top of the food chain, such as barracuda and groupers. Ciguatoxin is thermally stable and is not broken down by cooking. Symptoms can persist from weeks to years and include vomiting and hallucinations and, in extreme cases, allodynia, in which exposure to cold produces a painful burning sensation.

18 "Ladies make use of them to increase their milk": *The Gentleman's Magazine* (March 1753).

19 Background on the foundation, principles, diagnoses, and treatments in Traditional Chinese Medicine: Needham 1984; Maciocia 2005; Chan and Lee 2002; Kaptchuk 2000.

20 "One-third of the world's population relies on TCM for its health care": Von Moltke and Spaninks 2000.

21 For an account of the ancient and recent history of Chinese medicines, including "barefoot" doctors, see www.shen-nong.com.

22 Background on Chinese medicine in the twentieth century: Unschuld 1985; Fruehauf 1999.

23 "collected garbage of several thousand years": Unschuld 1985.

24 "uproot all shamanic beliefs and superstitions": Fruehauf 1999.

25 Discussion of animals using plants to self-medicate: Lozano 1998.

26 Malaria is caused by single-celled *Plasmodium*. The *falciparum* variety is the most dangerous: http://en.wikipedia.org/wiki/Plasmodium.

27 *Artemisia annua* appeared in the original *Shen Nong Ben Cao Jing*: Benksy and Gamble 1986.

28 The hunt for a traditional Chinese cure for malaria: Klayman 1985; BBC Horizon, *Defeating the Curse: How Science Is Tackling Malaria*. Program transcript, www.bbc.co.uk/sn/tvradio/programs/horizon.malaria.trans.shtml.

29 Discovery of artemisinin on the Potomac River: Klayman et al 1984.

30 Concerns over the current uses of artemisinin: Bate et al 2008; Anon 2008.

31 Tests on seahorse extracts: Benksy and Gamble 1986.

32 Genetic study of seahorses from Sun Yat-sen University in Guangzho: Zhang et al 2003. Researchers at Sun Yat-sen University have also identified a novel antimicrobial peptide in *Hippocampus kuda*: Wang et al 2008. The gene for the peptide has been inserted into yeast cells, and the resulting peptide was tested and found to be effective against two strains of bacteria, *Staphylococcus aureus* and *Staphylococcus saprophyticus* at low concentrations.

33 Another study claims that the high percentage of certain amino acids, fatty acids, and trace metals (e.g., magnesium, zinc, and iron) found in seahorses could make them a general health food, based on the same principle that many types of seafood are nutritious. However, individual seahorses are small and presumably large numbers would have to be consumed to gain health benefits comparable to a meal of regular fish or shellfish. The same vitamins are also readily available in tablet form: Lin et al 2008.

34 Peninsular Medical School study of tailor-made traditional remedies: Guo, Canter, and Ernst 2007.

35 The World Conservation Union (IUCN) enlists the help of expert biologists to assess the status of species that are thought to be threatened with extinction in the wild. Each species is assigned to a threat category ranging from "Critically Endangered," for those species facing an extremely high risk of extinction—the "red light" species—to the "Endangered," "Vulnerable," and "Near Threatened." Those classed as "Least Concern" are the "green light" species that are not currently considered to be adversely affected by human activities. Many species are listed as "Data Deficient" if not enough is known to fully assess their status: www.redlist.org.

36 The seahorse species listed as Vulnerable by IUCN are Barbour's, lined, Pacific, yellow, hedgehog, and three-spot.

37 Tonrentang was established in 1699 and is now one of the most valuable brands in China: Anon 2006, "Most valuable brands in China."

38 In the past, there was little demand for small, dark, spiny seahorses, which were deemed less therapeutic by ancient TCM texts, hence these seahorses were often left alone and not taken from the wild. Now that the demand is increasing for self-prescribed, so-called proprietary medicines, any seahorse, no matter how dark and spiny, can be used in the medicine trade, eliminating the former ad hoc protection of part of the seahorse family: Vincent, Marsden, and Sumaila 2007.

39 Composition and applications of "Seahorse genetic tonic pills": Zhu 1989.

40 For discussion of the environmental impacts of TCM, see Call 2006.

41 Survey of perceptions of U.S.-based TCM toward using replacements instead of endangered species: Call 2006.

42 Details of the plight of the saiga antelopes in Central Asia: Pearce 2003.

43 Discussion of the capacity for Viagra (sildenafil citrate) to reduce demand for traditional ingredients to boost male performance: von Hippel 1998; von Hippel and von Hippel 2002; Hoover 2003.

44 Report of high doses of lead in seahorse: McCormick 2000.

45 Estimated twenty-five million seahorses traded internationally: Project Seahorse Web site, http://seahorse.fisheries.ubc.ca/trade.html.

Chapter 4

1 Children swapping seahorses for sweets: Vincent 1996.

2 Small-scale seahorse fisheries in the Galápagos Islands, Mexi-

can, Central and South American trade: Baum and Vincent 2005.

3   Information on the Tanzanian seahorse fishery: McPherson and Vincent 2004.

4   Background on the Handumon Island seahorse fishery and Project Seahorse's work there: Pajaro et al 1997; Project Seahorse (Philippines), *2006 Annual Report*; Vincent et al 2007.

5   Background on destructive fishing techniques: Barber and Pratt 1998; Pet-Soede, Cesar, and Pet 2002.

6   Various terminology is used to describe areas of sea that are offered some level of protection from particular human activities, including fishing, dredging, and tourism: marine reserves, marine parks, and Marine Protected Areas (or MPAs) are all general terms, while No Take Zones (NTZs) or Highly Protected Marine Reserve (HPMRs) refer specifically to the strictest protection, where all damaging human activities are banned.

7   Study of coral and fish populations in Danajon Bank's marine reserves: Samoilys et al 2007.

8   Increase in seahorse size inside the Handumon reserve: Project Seahorse (Philippines), *2006 Annual Report*. In many fish species there is a direct relationship between body size and fecundity; larger individuals within a particular species are capable of producing more numerous, larger (and hence better surviving) eggs: Law 2001.

9   The use of trawlers to salvage remains of TWA flight 800: http://en.wikipedia.org/wiki/Trawling.

10 Background on the ecological and economic impacts of shrimp trawling: Environmental Justice Foundation 2003.

11 Lined seahorse bycatch in Florida: Baum, Meeuwig, and Vincent 2003.

12   Information and observations of the Cua Be trawl fishery in Vietnam were gathered during a visit to the area in September 2007. I am most grateful to Truong Si Ki and the staff at the Institute of Oceanography in Nha Trang for arranging my permits, facilitating my nighttime trip on a trawler, and for helping me to conduct interviews in Cua Be.

13   Average monthly wage determined from Gross National Income per capita, from country profile of Vietnam at www .worldbank.org.

14   Information on trade routes out of Vietnam: Giles et al 2005.

15   Tamil Nadu seahorse fisheries: Marichamy et al 1993.

16   Background on global seahorse trade including trade routes: Salin and Yohannan 2005; Project Seahorse Web site, http:// seahorse.fisheries.ubc.ca/trade; Vincent, Marsden, and Sumaila 2007.

17   Taiwanese seahorse import figures: Personal communication from TRAFFIC Taiwan.

18   Trade in seahorse curios in Latin America: Baum and Vincent 2005.

19   Background on the habitat damage caused by trawling: Environmental Justice Foundation 2003.

20   For a review of the biological basis for seahorses' vulnerability to exploitation: Foster and Vincent 2004.

21   Theory of K and r selected species: Begon, Harper, and Townsend 1996. Impact of trawling on K and r selected species: Environmental Justice Foundation 2003.

22   Shift in species in North West Australia due to trawl fisheries: Sainsbury 1998.

23   There are around eight thousand species of bryozoans (phylum Bryozoa), also known as moss animals or sea mats. They are tiny animals that live in colonies and often build stony

skeletons, a little like hard corals (family Scleractinia): International Bryozoology Association, http://nhm.ac.uk/hosted_sites/iba.

24 Experiments looking at the impact of beach seining on two seahorse species in Ria Formosa lagoon, Portugal: Curtis et al 2007.

25 Fishermen's and traders' perceptions of declining seahorse catches and abundance: Giles et al 2005; Vincent 1996; Baum and Vincent 2005; Marichamy et al 1993.

26 "Water-diamonds" paradox and examples of rare species and specimens being more highly valued: Courchamp et al 2006.

27 Exploitation of orange roughy, *Hoplostethus atlanticus:* Clark et al 2000.

28 Near extinction of barndoor skates, *Dipturus laevis:* Casey and Myers 1998.

29 The barndoor skate is listed as "Endangered" by IUCN rather than the more urgent classification of "Critically Endangered" because fishing effort in their range has declined in the last decade, their range is significantly wider than previously thought, and the number of juveniles now appears to be increasing, especially in No Take Zones on Georges Bank and southern New England shelf: www.iucnredlist.org/search/details.php/39771/summ.

30 Background information on CITES: www.cites.org.

31 Amanda Vincent's global assessment of the seahorse trade: Vincent 1996.

32 Listing of seahorses on CITES: Koldewey 2005.

33 The word *sustainable* is often used loosely in the realm of conservation and development without definition of what the term specifically means. Sustainable over what time frame and for whom: individuals, private companies, governments,

seahorses? Within fisheries management, the term is generally intended to refer to a fishery that operates only within certain boundaries and leaves behind enough individuals to maintain the wild population, and which does not cause noticeable declines in exploited stocks. One problem is that the "sustainable" catch size can be very difficult to predict—and even harder to enforce—and can change annually with natural fluctuations in environmental factors that influence population recruitment and survival.

34 Temporary ten-cm size limit on traded seahorses: Foster and Vincent 2005. A theoretical model of seahorse population dynamics in 2008 indicated that the ten-centimeter size limit is unlikely to prevent population declines in many species due to insufficient breeding opportunities for seahorses before they become vulnerable to exploitation in fisheries: Curtis and Vincent 2008. It is suggested that an increase in the minimum size limit to at least fourteen centimeters would allow fishermen to retain ninety percent of their earnings, while at the same time halving the risk of decline and extinction in exploited populations of seahorses.

35 Study of seahorses on sale in Californian curio shops and Chinese pharmacies: Sanders et al 2008.

36 Illegal status of seahorse fisheries in the Philippines: Project Seahorse (Philippines), *2006 Annual Report.*

37 Ninety-nine percent survival of seahorses in Hernando Beach trawl fishery: Baum, Meeuwig and Vincent 2003.

38 Low survival of Tampa Bay dwarf seahorses: Baum, Meeuwig and Vincent 2003.

39 Since the 1980s, it has been mandatory for all shrimp trawlers operating in North American waters to fix Turtle Excluder Devices, or TEDs, into their nets. The same now goes for foreign fleets that export shrimp to the United States, follow-

ing the introduction of the shrimp turtle law in 1989. This might appear to be a win-win situation, but many fishermen resent having to use these devices; they don't come cheap, and fishers believe they let not only turtles and dolphins slip through but also much of the valuable catch. A study in Texas found that over forty percent of shrimpers were tying up their TEDs, making them ineffective at excluding air-breathing megafauna: Environmental Justice Foundation 2003.

Background on bycatch reduction devices: NSW Department of Primary Industries, *Bycatch and its reduction*: www .fisheries.nsw.gov.au/commercial/commercial2/bycatch_ and_its_reduction.

Chapter 5

1 Background on the history of aquarium keeping: Brunner 2003.
2 "To a person not instructed in natural history": Huxley 1854.
3 Details of Victorian craze for natural history: Merrill 1989.
4 The English counties of Cornwall, Devon, Dorset, and Somerset are known as the West Country.
5 Background on Philip Henry Gosse: Thwaite 2002.
6 Invention of the Wardian Case: Orlean 1998.
7 The stony corals (order Scleractinia) that Anna Thynne kept were solitary polyps such as the Devonshire cup-coral, *Caryophyllia smithii*.
8 Biography of Anna Thynne: Stott 2003.
9 Warington's presentation to the Chemical Society in London: Brunner 2003.
10 "ransack the world": Lewes 1862.
11 "England has become Gosse-ified": *The Atlas* (25 October 1856).

12   William Alford Lloyd's Aquarium Warehouse: Brunner 2003.

13   "The wonders of the ocean do not reveal themselves to vulgar eyes": Humphreys 1857.

14   "hardly necessary to recommend": Sowerby 1857.

15   "a geological hammer": Hibberd 1856.

16   Gosse setting up the first public aquarium at London Zoo: Thwaite 2002; Blunt 1976.

17   "so crowded daily with its curious visitors": Humphreys 1857.

18   Barnum bringing aquariums to the United States: Brunner 2003; Butler 1858.

19   Dropsy or edema in aquarium fish is a common disease usually caused by a bacterial infection: http://en.wikipedia.org/wiki/Fish_Dropsy.

20   Mr. Pinto bringing seahorses to London from Portugal: Brightwell 1936.

21   "They will entwine their tails together": *Glasgow Herald* (1866).

22   "the hero of the Aquarium": *Freeman's Journal and Daily Commercial Advertiser* (1873).

23   "prickly manes, and a motor in the idle of their backs": *Daily Telegraph*, London (1869).

24   Report of seahorse births at Manchester Aquarium: *Manchester Courier* (1873).

25   "most people have heard of, and may have seen": *Boys Own Paper* (27 August 1898).

26   "See how proud Mr. Hip. appears to be": *Kind Words for Boys and Girls* (25 March 1869).

27   "enough to mount a whole regiment of sea fairies": *Little Folks* magazine (1873).

28   "This decoration falls off as the fish gets into deep water": *The Ladies' Treasury: A Household Magazine* (1 June 1892).

29   Westminster aquarium poem: *Funny Folks* (23 October 1875).

30 "Nine times out of ten they take to torturing something":
    Collins 1868.

31 "when the mode of transporting delicate creatures is better
    understood": Holdsworth 1860.

32 "as yet, a plaything, a mere toy": Humphreys 1857.

33 Following the original popularity of "Seahorses: Beyond
    Imagination" at the John G. Shedd Aquarium in Chicago, the
    exhibition moved to various other aquariums around the world
    and has been open permanently at the Tennessee Aquarium
    since 2002: www.tnaqua.org/Seahorses/Seahorse_home.asp.

34 I am indebted to Karolina Alford for showing me around the
    Tennessee Aquarium and answering my many questions
    about syngnathid keeping.

35 Information on seahorse husbandry: Koldewey 2005.

36 For some reason, captive seadragons do not like to be in the
    pitch-dark and panic if their night lights fail, floundering
    at the water surface.

37 Captive breeding at Birch Aquarium, La Jolla, California:
    http://aquarium.ucsd.edu/Education/Learning_Resources/
    Secrets_of_the Seahorse.

38 "Suddenly Kokino gave a little grunt of half surprise": Dur-
    rell 1969.

39 I gained great insights into the trials and tribulations of sea-
    horse keeping from Carmine Scotch, Deanne Ward, and
    David Warland at Syngnathid.org, Kells at Seahorsesuk.net,
    and the online forums at Marine Fish UK and Seahorse.com.

40 Details of seahorse pathology: Berzins and Greenwell 2005.

41 Survival of seadragons after Hurricane Katrina: Anon 2005,
    "Katrina kills most fish in New Orleans aquarium." Shortly
    after the disaster, the seadragons, as well as the surviving sea
    otters and penguins, were safely airlifted out and taken to
    other aquariums and zoos around the United States.

42 Early Chinese attempts to farm seahorses: Forteath 2000.

43 I am most grateful to Rachelle Hawkins from Seahorse Australia for providing me an insight into the practicalities of seahorse farming.

44 Report of seahorses flown from Paris to London in the 1920s: Anon 1924, "Flying Fish."

45 Seahorses seized at Stansted Airport: Burleigh 2006.

46 Details of Ocean Rider's environmental policy: www.seahorse .com/Ocean_Rider_Conservation.

47 There are Web sites offering West African seahorses for sale from ranches in Cameroon, both alive for the aquarium trade and dead for the medicine trade.

48 Overview of wildlife farming: Damania and Bulte 2007.

49 On bear farming: Mills, Chan, and Ishihara 1995.

50 Captive-bred fish are less fit for survival in the wild: Araki, Cooper, and Blouin 2007.

51 I am indebted to David Harasti for providing information on the Australian seahorse release program.

52 Background on surgeon John White: http://en.wikipedia .org/wiki/John_White_(surgeon).

53 "This animal, like the Flying Fish, being commonly known": White 1790. White's book is freely available online at Project Guttenberg Australia, http://gutenberg.net.au/ebooks03/ 0301531h.html

54 "capacity of amateurs": Thwaite 2002.

Chapter 6

1 *The Life Aquatic with Steve Zissou*, 2004. Touchstone Pictures. Director Wes Anderson. Writers Wes Anderson, Noah Baumbach. Starring Bill Murray, Owen Wilson, Cate Blanchett, Anjelica Huston, Willem Dafoe, Jeff Goldblum, Michael Gambon.

2 "How could you! You horrible boy!": Arthur, 1964.

3 *The Little Mermaid* is based on a story by Hans Christian Andersen.

4 *The Little Mermaid: Ariel's Undersea Adventures—Stormy the Wild Seahorse*, 1993. Walt Disney Video. Director Jamie Mitchell. Starring Bradley Pierce, Joachim Kemmer.

5 Details of *Pokémon*: http://en.wikipedia.org/wiki/Pok %C3A9mon.

6 The inspiration to create an animal-collecting game apparently came from *Pokémon* creator Satoshi Tajiri's love of insect collecting.

7 A study from 2002 found that British schoolchildren could identify eighty percent of a random sample of *Pokémon* "species" but less than fifty percent of species of British wildlife, highlighting young children's innate capacity for learning about creatures, even made-up ones, and the growing detachment of urban societies to native wildlife: Balmford et al 2002.

8 *SpongeBob SquarePants*, "My Pretty Seahorse," Season 3, Episode 12, 2002. Nickelodeon. Writers Kent Osborne, Derek Drymon, Stephen Hillenburg, Mark O'Hare. Starring Tom Kenny, Rodger Bumpass, Frank Welker.

9 *Finding Nemo*, 2003. Walt Disney Pictures. Directors Andrew Stanton, Lee Unkrich. Writers Andrew Stanton, Bob Peterson, David Reynolds. Starring Ellen DeGeneres, Albert Brooks, Geoffrey Rush, Willem Dafoe.

10 *Shark Tale*, 2004. Dreamworks Animation. Directors Bibo Bergeron, Vicky Jenson, Rob Letterman. Writers Michael J. Wilson, Rob Letterman. Starring Will Smith, Robert De Niro, Renée Zellweger, Jack Black.

11 NME interview with John Squire: Udo 1996.

12 Details of the history of underwater photography: Adam 1993.

13 Background on Jean Painlevé and his films: Bellows, Mc-Dougall, and Berg 2000.

14 Painlevé's 1934 film *L'Hippocampe ou Cheval Marin* is available on DVD, together with a collection of his other films: *Science Is Fiction: The films of Jean Painlevé/The Sounds of Science, 1927–1978*, produced by the British Film Institute.

15 "Everything about this animal": Painlevé, *The Seahorse*, quoted in Bellows, McDougall, and Berg 2000.

16 Stories of the early trials of scuba gear: Cousteau, Dumas, and Dugan 1973.

17 The exact number of certified divers around the world is difficult to pin down and often overinflated, but there could be as many as a million people in the United States and six million people across Europe: Davison 2007; World Recreational Scuba Training Council, *Facts and Figures*.

18 Details of Bonaire seahorses and study of coral damage from seahorse spotters: Uyarra and Coté 2007.

19 Observations of seahorse tourism in Belize were made during a visit to Ambergris Cay in March 2008.

20 David Harasti provided me with details of the tagging study of White's seahorses in Port Stephens.

21 Mediterranean Hippocampus Mission: Goffredo, Piccinetti, and Zaccanti 2004.

22 Studies have shown that both short- and long-snouted seahorses can have dermal fronds (manes) so this may not be a reliable characteristic to distinguish between them: Curtis 2006.

23 Following the success of the Mediterranean Hippocampus Mission, a new venture has been established called Diving for Environment: Mediterranean Underwater Biodiversity Project, which monitors seahorses and fifty-nine other marine species.

24 For background on the latest work of The Seahorse Trust and the British Seahorse Survey: Garrick-Maidment 2007; The Seahorse Trust Web site: www.theseahorsetrust.co.uk; The British Seahorse Survey Web site: http://britishseahorsesurvey.org.

25 Reports of seadragon sightings are now being collected by ReefWatch in Australia: www.reefwatch.asn.au.

26 I am grateful to Denise Tackett for telling me her story of the pygmy seahorses.

27 Taxonomic description of Denise's pygmy seahorse: Lourie and Randall 2003. Description of Coleman's pygmy seahorse: Kuiter 2003. Description of three Indonesian pygmy seahorses: Lourie and Kuiter 2008. A potential new species of seahorse from the Red Sea was found by underwater photographer Helmut Debelius after he put out a call in a dive magazine for anyone who has seen a similar seahorse. The lack of reported sightings confirmed the rarity of the newly described species, *Hippocampus debelius*.

28 Resources for seahorse identification: www.fishbase.org; Lourie et al 2004; Kuiter 2009. (N.B.: Many of the species listed by Kuiter are considered by other experts to be synonyms.)

29 Population biology and ecology of the Cape seahorse, and death of three thousand of them in 1991: Bell et al 2003; Lockyear et al 2006.

30 Warming seas around British coasts and changes in ocean ecosystems: Beaugrand et al 2002. Planktonic changes in the North Sea may explain the sudden increase in abundance of oceangoing snake pipefish (*Entelurus aequoreus*) in the North Sea and northeast Atlantic in recent years: Harris et al 2007.

31 Background on the impacts of climate change on coral reefs: Hoegh-Guldberg et al 2007. Over the past two hundred years, emissions of carbon dioxide from human activities have al-

ready led to a reduction in the average pH of surface seawater of 0.1 units and could fall by 0.5 units by the year 2100: Royal Society policy statement on ocean acidification: http://royalsociety.org/document.asp?id=3249. The problem of coral bleaching is not only exacerbated by steadily increasing sea temperatures but also triggered by short-lived temperature anomalies linked to the El Niño–Southern Oscillation in the Pacific Ocean. Since the 1980s, the apparent increase in the frequency, persistence, and intensity of El Niño events is thought by some to be linked to climate change: National Oceanic and Atmospheric Administration (NOAA), *What is an El Niño?* www. pmel.noaa.gov/elnino/el-nino-story.html. The worst global bleaching event on record so far was in 1998, which killed sixteen percent of the world's coral reefs, including half of the reefs in the Indian Ocean: Wilkinson 2004.

32 Melting soft corals: Milstein 2007.

33 Importance of marine reserves for promoting resilience to climate change: Black 2008.

34 Convention on Biological Diversity: www.cbd.int.

35 The Sea Around Us project, in collaboration with WWF, UNEP WCMC, and the IUCN have established an online database of MPAs around the world: www.mpaglobal.org. In 2005, Louisa Wood announced at the first international MPA congress that the current protection of the oceans added up to around one percent.

36 I am grateful to Choo Chee Kuang for providing information on the SOS Malaysia project. For more information see www.sosmalaysia.org.

37 Reports of seahorses in the Thames: Leake and Waite 2008.

38 The two species of British seahorse, the long-snouted and short-snouted seahorse, are now both protected under the

Wildlife and Countryside Act 1981, making it illegal to deliberately collect or damage them or their habitat.

39 Death of the Thames whale: Knight 2006.

Epilogue

1 Background on keystone species: Begon, Harper, and Townsend 1996.

2 Ecosystem impacts of sea otter hunting: Jackson et al 2001.

3 "The overwhelming reason is man's imaginative health": Kingsnorth 2001.

BIBLIOGRAPHY

Adam V. William Thompson—100 years of underwater photography? British Society of Underwater Photographers newsletter *In Focus* (September 1993).

Aelian. *De Natura Animalium*. Trans. Scholfield AF. London: Heinemann, 1958.

Allen JR, Anderson J. *The Early Christian Monuments of Scotland: A Classified, Illustrated, Descriptive List of the Monuments, with an Analysis of Their Symbolism and Ornamentation*. 2 vols. Edinburgh: Society of Antiquaries of Scotland/Neill & Co., 1903.

Anon. Flying fish, seahorses' trip by air to the zoo aquarium. *The Times* (London) (May 1924).

Anon. Island seahorse drifts off course. *BBC News* (17 March 2006). http://news.bbc.co.uk/1/hi/world/europe/jersey/4818358.stm.

Anon. Katrina kills most fish in New Orleans aquarium. *CNN .com* (9 September 2005). www.cnn.com/2005/TECH/science/09/07/katrina.zoos.

Anon. Most valuable brands in China. *China Daily* (6 June 2006). www.chinadaily.com.cn/bizchina/2006-06/06/content_609744.htm.

Anon. Resisting arrest. Fighting malarial drug resistance. *The Economist* (17 May 2008).

Anon. Ten charged in missing brooch case. *Turkish Daily News* (14 July 2006).

Araki H, Cooper B, Blouin MS. Genetic effects of captive breeding cause a rapid, cumulative fitness decline in the wild. *Science* 2007, 318:100–103.

Aristotle. *On the History of Animals.* Trans. Wentworth Thompson D. eBooks@Adelaide 2007. http://ebooks.adelaide.edu.au/a/aristotle/history.

Arsu S. Thefts focus attention on lax security at Turkey's museums. *The New York Times* (13 June 2006).

Arthur RM. *Carolina and the Sea-Horse: And Other Stories.* London: George G. Harrap and Co. Ltd, 1964.

Ashley-Ross MA. Mechanical Properties of the Dorsal Fin Muscle of Seahorse (*Hippocampus*) and Pipefish (*Syngnathus*). *Journal of Experimental Zoology* 2002, 293:561–577.

Attenborough, D. *Life in Cold Blood.* BBC documentary 2007.

Balmford A, Clegg L, Coulson T, Taylor J. Why conservationists should heed Pokémon. *Science* 2002, 295:2367.

Barber CV, Pratt VR. Poison and profits. *Environment* 1998, 40:1–12.

Bate R, Coticelli P, Tren R, Attaran A. Antimalarial drug quality in the most severely malarious parts of Africa—a six country study. *PLoS One* (7 May 2008). www.plosone.org/article/info:doi/10.1371/journal.pone.0002132.

Baum JK, Meeuwig JJ, Vincent ACJ. Bycatch of lined seahorses (*Hippocampus erectus*) in a Gulf of Mexico shrimp trawl fishery. *Fishery Bulletin* 2003, 101:721–731.

Baum JK, Vincent ACJ. Magnitude and inferred impacts of the seahorse trade in Latin America. *Environmental Conservation* 2005, 32:305–319.

Beaugrand G, Reid PC, Ibanez F, Lindley JA, Edwards M. Reorganization of North Atlantic marine copepod biodiversity and climate. *Science* 2002, 296:1692–1694.

Begon M, Harper JL, Townsend CR. *Ecology: Individuals, Populations and Communities.* Oxford: Wiley Blackwell, 1996.

Bell EM, Lockyear JF, McPherson JM, Marsden AD, Vincent ACJ. First field studies of an endangered South African seahorse, *Hippocampus capensis. Environmental Biology of Fishes* 2003, 67:35–46.

Bellows AM, McDougall M, Berg B (eds.). *Science Is Fiction, The Films of Jean Painlevé.* San Francisco: Brico Press, 2000.

Bennett G. *Wanderings in New South Wales.* London: Richard Bentley, 1834.

Bensky D, Gamble A. *Chinese Herbal Medicine: Materia Medica.* Seattle: Eastland Press, 1986.

Berzins IK, Greenwell MG. Syngnathid health management. In *Syngnathid Husbandry in Public Aquariums.* Ed. Koldewey H. Project Seahorse and the Zoological Society of London, 2005: 28–38.

Black R. Fish key to reef climate survival. *BBC News* (20 March 2008). http://news.bbc.co.uk/l/hi/sci/tech/7306693.stm.

Blunt W. *The Ark in the Park: The Zoo in the Nineteenth Century.* London: Hamish Hamilton Ltd, 1976.

Briggs KM. *Dictionary of Fairies: Hobgoblins, Brownies, Bogies and Other Supernatural Creatures.* London: Allen Lane, 1977.

Brightwell LR. *The Zoo You Knew.* Oxford: Blackwell, 1936.

Brunner B. *The Ocean at Home: An Illustrated History of the Aquarium.* New York: Princeton Architectural Press, 2003.

Burleigh J. Rare seahorses survive Atlantic journey in parcel. *The Independent* (UK) (25 February 2006). www.independent.co.uk/news/uk/this-britain/rare-seahorses-survive-atlantic-journey-in-parcel-467638.html.

Butler HD. *The Family Aquarium; or, Aqua Vivarium: A "New Pleasure" for the Domestic Circle: Being a Familiar and Complete Instructor upon the Subject of the Construction, Fitting-up, Stocking, and Maintenance of the Fluvial and Marine Aquaria, or "River and Ocean Gardens."* New York: Dick & Fitzgerald, 1858.

Call E. *Mending the Web of Life: Chinese Medicine and Species Conservation.* Gaithersburg, MD: Signature Book Printing, Inc., 2006.

Carson R. *The Edge of the Sea.* London: Staple Press, 1955.

Casey JM, Myers RA. Near extinction of a large, widely distributed fish. *Science* 1998, 281:690–692.

Casey SP, Hall HJ, Stanley HF, Vincent ACJ. The origin and evolution of seahorses (genus *Hippocampus*): A phylogenetic study using the cytochrome b gene of mitochondrial DNA. *Molecular Phylogenetics and Evolution* 2004, 30:261–272.

Cessford C. Pictish art and the sea. *The Heroic Age: A Journal of Early Medieval Northwestern Europe* 2005, 8. www.mun.ca./mst/heroicage/issues/8/cessford.html.

Chan K, Lee H, eds. *The Way Forward for Chinese Medicine.* London and New York: Taylor and Francis, 2002.

Choo CK, Liew HC. Exploitation and trade in seahorses in Peninsular Malaysia. *Malayan Nature Journal* 2005, 57:57–66.

Clark MR, Anderson OF, Francis R, Tracey DM. The effects of commercial exploitation on orange roughy (*Hoplostethus atlanticus*) from the continental slope of the Chatham Rise, New Zealand, from 1979 to 1997. *Fisheries Research* 2000, 45:217–238.

Clynick BG. Harbour swimming nets: a novel habitat for seahorses. *Aquatic Conservation: Marine and Freshwater Ecosystems* 2008, 18:483–492.

Collins W. *The Moonstone.* New York: Signet Classic, Penguin Books, 1868.

Colson DJ, Patek SH, Brainerd EL, Lewis SM. Sound production during feeding in *Hippocampus* seahorses (Syngnathidae). *Environmental Biology of Fishes* 1998, 51:221–229.

Courchamp F, Angulo E, Rivalan P, Hall RJ, Signoret L, Bull L, Meinard Y. Rarity value and species extinction: The anthropogenic allee effect. *PloS Biology* 2006, 4:415.

Cousteau J. *The Ocean World.* New York: Abradale Press/Harry N. Abrams, Inc., 1979.

Cousteau J, Dumas F, Dugan J. *The Silent World.* Toronto: Balantine Books, 1973.

Curtis JMR. A case of mistaken identity: Skin filaments are unreliable for identifying *Hippocampus guttulatus* and *Hippocampus hippocampus. Journal of Fish Biology* 2006, 69:1855–1859.

Curtis JMR, Vincent ACJ. Use of Population Viability Analysis to Evaluate CITES Trade-Management Options for Threatened Marine Fish. *Conservation Biology* 2008, 22:1225–1232.

Curtis JMR, Ribeiro J, Erzini K, Vincent ACJ. A conservation trade-off? Interspecific differences in seahorse responses to experimental changes in fishing effort. *Aquatic Conservation: Marine and Freshwater Ecosystems* 2007, 17:468–484.

Dahl R. "The Mildenhall Treasure." In *The Wonderful Story of Henry Sugar and Six More*. London: Jonathon Cape, 1977.

Damania R, Bulte EH. The economics of wildlife farming and endangered species conservation. *Ecological Economics* 2007, 62:461–472.

Davison B. How many divers are there? *Undercurrent Scuba Diving Magazine* (May 2007).

Durrell G. *Birds, Beasts and Relatives*. London: Collins, 1969.

Eastman CR. Olden time knowledge of Hippocampus. *Smithsonian Institution Annual Report* 1915:349–357.

Environmental Justice Foundation (EJF). *Squandering the Seas: How Shrimp Trawling Is Threatening Ecological Integrity and Food Security Around the World*. London: Environmental Justice Foundation, 2003.

Erdem S. Curator held after tip-off on Croesus' stolen riches. *The Times* (London) (20 May 2006).

Forteath N. Farmed seahorses: A boon to the aquarium trade. *INFOFISH International* 2000, 3:48–50.

Foster SJ, Vincent ACJ. Enhancing sustainability of the international trade in seahorses with a single maximum size limit. *Conservation Biology* 2005, 19:1044–1050.

Foster SJ, Vincent ACJ. Life history and ecology of seahorses: Implications for conservation and management. *Journal of Fish Biology* 2004, 65:1–61.

Fruehauf H. Science, politics and the making of "TCM": Chinese medicine in crisis. *Journal of Chinese Medicine* (October 1999). www.jcm.co.uk/media/sample_articles/tcmcrisis.pdf.

Garrick-Maidment N. *British Seahorse Survey 2007*. The Seahorse Trust, 2007.

Giles BG, Si Ky T, Huu Hoan D, Vincent ACJ. The catch and trade of seahorses in Vietnam. *Biodiversity and Conservation* 2005 15: 2497–2513.

Gill T. The life history of the sea-horses (Hippocampids). *Proceedings of the United States National Museum* 1905, 28:805–814.

Goffredo S, Piccinetti C, Zaccanti F. Volunteers in marine conservation monitoring: A study of the distribution of seahorses carried out in collaboration with recreational scuba divers. *Conservation Biology* 2004, 18:1492–1503.

Gomon MF, Kuiter RH. Two new pygmy seahorses (Teleostei: Syngathidae: *Hippocampus*) from the Indo-West Pacific. *Aqua, International Journal of Ichthyology* 2009, 15: 37–44.

Gosse PH. *A Naturalist's Rambles on the Devonshire Coast*. London: John van Voorst, 1853.

Gosse PH. *The Aquarium: An Unveiling of the Wonders of the Deep Sea*. London: John van Voorst, 1854.

Griggs B. *Green Pharmacy: A History of Herbal Medicine*. London: Jill Norman and Hobhouse Ltd, 1981.

Guo R, Canter PH, Ernst E. A systematic review of randomised clinical trials of individualised herbal medicine in any indication. *Postgraduate Medical Journal* 2007, 83:633–637.

Handwick B, Hafvenstein L. Belize reef die-off due to climate change? *National Geographic News* (25 March 2003). http://news.nationalgeographic.com/news/2003/03/0325_030325_belizereefs.html.

Hard R. *The Routledge Handbook of Greek Mythology.* London and New York: Routledge, 2004.

Harris MP, Beare D, Toresen R, Nottestad L, Kloppmann M, Dorner H, Peach K, Rushton DRA, Foster-Smith J, Wanless S. A major increase in snake pipefish (*Entelurus aequoreus*) in northern European seas since 2003: Potential implications for seabird breeding success. *Marine Biology* 2007, 151:973–983.

Henderson G, Henderson I. *The Art of the Picts: Sculpture and Metalwork in Early Medieval Scotland.* London: Thames and Hudson Ltd, 2004.

Henderson I. Pictish Monsters: Symbol, text and image. In *Chadwick Memorial Lectures.* Cambridge: Department of Anglo-Saxon, Norse and Celtic, University of Cambridge, 1997.

Hibberd JS. *Rustic Adornments for Homes of Taste, and Recreations for Town Folk, in the Study and Imitation of Nature.* London, 1856.

Hobbs R. "The Mildenhall Treasure": Roald Dahl's ultimate tale of the unexpected. *Antiquity* 1997, 71:63–73.

Hoegh-Guldberg O, Mumby PJ, Hooten AJ, Steneck RS, Greenfield P, Gomez E, Harvell CD, Sale PF, Edwards AJ, Caldeira K. Coral reefs under rapid climate change and ocean acidification. *Science* 2007, 318:1737–1742.

Holdsworth EWH. *Hand-Book to the Fish-House in the Gardens of the Zoological Society.* London, 1860.

Holmquist JG. Benthic macroalgae as a dispersal mechanism for fauna: Influence of a marine tumbleweed. *Journal of Experimental Marine Biology and Ecology* 1994, 180:235–251.

Homer. *The Iliad.* Trans. Fagles R. New York: Penguin Classics, 1998.

Homer. *The Odyssey.* Trans. Fagles R. New York: Penguin Classics, 2006.

Hoover C. Response to "Sex, drugs and animals parts: Will Viagra save threatened species?" *Environmental Conservation* 2003, 30:317–318.

Hoving T. *Making the Mummies Dance: Inside the Metropolitan Museum of Art.* New York: Simon & Schuster, 1993.

Howel MO. *The Horse in Magic and Myth.* Mineola, NY: Dover Publications, 2002.

Humphreys HN. *Ocean Gardens: The History of the Marine Aquarium, and the Best Methods Now Adopted for Its Establishment and Preservation.* London: Sampson Low, Son and Co., 1857.

Huxley TH. *On the Educational Value of the Natural History Sciences.* London, 1854.

Isaacs J. *Australian Dreaming: 40,000 years of Aboriginal History.* London: New Holland Publishers, 2005.

Jackson A. *Pictish Symbol Stones?* Edinburgh: Association for Scottish Ethnography, University of Edinburgh, 1993.

Jackson JBC, Kirby MX, Berger WH, Bjorndal KA, Botsford LW, Bourque BJ, Bradbury RH, Cooke R, Erlandson J,

Estes JA. Historical overfishing and the recent collapse of coastal ecosystems. *Science* 2001, 293: 629–638.

Kaptchuk T. *The Web That Has No Weaver: Understanding Chinese Medicine.* London: Rider, 2000.

Kingsnorth P. Strife on Earth. *The Ecologist* (April 2001).

Klayman DL. Qinghaosu (artemisinin): An antimalarial drug from China. *Science* 1985, 228:1049–1055.

Klayman DL, Lin AJ, Acton N, Scovill JP, Hoch JM, Milhous WK, Theoharides AD, Dobek AS. Isolation of artemisinin (qinghaosu) from *Artemisia annua* growing in the United States. *Journal of Natural Products* 1984, 47:715–717.

Knight S. Thames whale dies of thirst, muscle and kidney damage. *Times Online* (London) (25 January 2006). www.timesonline.co.uk/tol/news/uk/article719555.ece.

Koldewey H. Seahorses take to the world stage. *Live Reef Fish Information Bulletin* 2005, 13:33–34.

Koldewey H, ed. *Syngnathid husbandry in public aquariums, 2005 Manual.* Zoological Society of London and Project Seahorse, 2005.

Kuiter RH. A new pygmy seahorse (Pisces: Syngnathidae: *Hippocampus*) from Lord Howe Island. *Records of the Australian Museum* 2003, 55:113–116.

Kuiter RH. Revision of the Australian seahorses of the genus *Hippocampus* (Syngnathiformes: Syngnathidae) with descriptions of nine new species. *Records of the Australian Museum* 2001, 53:293–340.

Kuiter RH. *Seahorses, Pipefishes and Their Relatives.* Aquatic Photographics, 2009.

Laidler L, Laidler K. *China's Threatened Wildlife*. London: Blandford, 1996.

Lavergne D, Ashmore PJ. *Aberlemno Pictish Carved Stones*. Historic Scotland, 2001.

Law, R. Phenotypic and genetic changes due to selective exploitation. In *Conservation of Exploited Species*. Eds. Reynolds JD, Mace GM, Redford KH, Robinson JG. Cambridge University Press, 2001, 323–343.

Leake J, Waite R. And they're off: seahorses reach the Thames. *The Sunday Times* (London) (6 April 2008).

Lewes GH. *Studies in Animal Life*. London: Smith, Elder and Co., 1862.

Lin Q, Lin J, Lu J, Li B. Biochemical composition of six seahorse species, *Hippocampus* sp., from the Chinese coast. *Journal of the World Aquaculture Society* 2008, 39:225–234.

Linnaeus C. *Systema Naturae*. Holmiae: Impensis, 1758.

Lloyd S. *Ancient Turkey. A Traveler's History of Anatolia*. London: British Museum Publications, 1989.

Lockyear JF, Hecht T, Kaiser H, Teske PR. The distribution and abundance of the endangered Knysna seahorse *Hippocampus capensis* (Pisces: Syngnathidae) in South African estuaries. *African Journal of Aquatic Sciences* 2006, 31:275–283.

Lourie SA, Foster SJ, Cooper EWT, Vincent ACJ. *A Guide to the Identification of Seahorses*. Washington, DC: TRAFFIC North America, 2004. http://seahorse.fisheries.ubc.ca/IDguide.html.

Lourie SA, Kuiter RH. Three new pygmy seahorse species from Indonesia (Teleostei: Syngnathidae: *Hippocampus*). *Zootaxa*

2008, 1963: 54–68. www.mapress.com/zootaxa/2008/f/
zto1963p068.pdf.

Lourie SA, Pritchard JC, Casey SP, Truong SK, Hall HJ, Vincent ACJ. The taxonomy of Vietnam's exploited sea-horses (family Syngnathidae). *Biological Journal of the Linnean Society* 1999, 66:213–256.

Lourie SA, Randall JE. A new pygmy seahorse, *Hippocampus denise* (Teleostei: Syngnathidae), from the Indo-Pacific. *Zoological Studies* 2003, 42:284–291.

Lozano GA. Parasitic stress and self-medication in wild animals. *Advances in the Study of Behavior* 1998, 27:291–317.

Lussanet MHE, Muller M. The smaller your mouth, the longer your snout: Predicting the snout length of *Syngnathus acus, Centriscus scutatus* and other pipette feeders. *Journal of the Royal Society Interface* 2007, 4:561–573.

Maciocia G. *The Foundations of Chinese Medicine: A Comprehensive Text for Acupuncturists and Herbalists.* Edinburgh: Elsevier Churchill Livingstone, 2005.

Mack A. *Field Guide to the Pictish Symbol Stones.* Scotland: Pinkfoot Press, 1997.

Maguire EA, Gadian DG, Johnsrude IR, Good CD, Ashburner J, Frackowiak RSJ, Frith CD. Navigation-related structural changes in the hippocampi of taxi drivers. *Proceedings of the National Academy of Sciences of the United States of America* 2000, 87:4398–4403.

Marichamy R, Lipton AP, Ganapathy A, Ramalingam JR. Large scale exploitation of seahorse (*Hippocampus kuda*) along the Palk Bay coast of Tamil Nadu. *Marine Fisheries Information Service* 1993, 119:17–20.

Martin-Smith KM, Lam TF, Lee SK. Trade in pipehorses *Solegnathus* sp. for Traditional Medicine in Hong Kong. *TRAFFIC Bulletin* 2003, 19:139–148.

Martin-Smith KM, Vincent ACJ. Exploitation and trade of Australian seahorses, pipehorses, seadragons and pipefishes (family Syngnathidae). *Oryx* 2006, 40:141–151.

Matthioli PA. *Commentarii in sex libros Pedacii Dioscoridis Anazarbei de medica materia.* 1565.

McCormick E. Herbs can hurt instead of heal. State officials warn some Asian remedies also may be toxic. *San Francisco Chronicle* (12 March 2000). www.sfgate.com/cgi-bin/article .cgi?file=/examiner/archive/2000/03/12/NEWS13377. dtl&type=health.

McPherson JM, Vincent ACJ. Assessing East Africa trade in seahorse species as a basis for conservation under international controls. *Aquatic Conservation: Marine and Freshwater Ecosystems* 2004, 14:521–538.

Merrill LL. *The Romance of Victorian Natural History.* Oxford: Oxford University Press, 1989.

Mills JA, Chan S, Ishihara A. *The Bear Facts: The East Asian Market for Bear Gall Bladder.* Cambridge: TRAFFIC Network Report, 1995.

Milstein M. Soft corals "melting" due to warming seas, expert says. *National Geographic News* (13 July 2007). http://news .nationalgeographic.com/news/2007/07/070713-corals-melt .html.

Morford MPO, Lenardon RJ. *Classical Mythology.* Oxford: Oxford University Press, 2007.

Murray MA. An Egyptian Hippocampus. In *Historical Studies.*

Eds. Knobel EB, Midgley WW, Milne JG, Murray MA, Petrie WMF. Egypt: London School of Archaeology, 1911.

Needham J. *Science and Civilisation in China*. Vol. 6 of *Biology and Biological Technology*. Cambridge: Cambridge University Press, 1984.

Norman G. Talking Turkey: Who owns the treasures of antiquity? *The Independent* (UK) (13 June 1993).

Norman G. A confusing case of losing your marbles: Turkey is fighting to retrieve from American collections some notable antiquities it believes were smuggled abroad. *The Independent* (UK) (2 February 1991).

Nowak R. Plenty more fish in the sea. *New Scientist* (19 May 2007).

Orlean S. *The Orchid Thief: A True Story of Beauty and Obsession*. London: Random House, 1998.

Pajaro MG, Vincent ACJ, Buhat DY, Perante NC. The role of seahorse fishers in conservation and management. In *First International Conference in Marine Conservation, Hong Kong*, 1997:118–126.

Pearce F. Rhino rescue plan decimates Asian antelopes. *New Scientist* (12 February 2003).

Pet-Soede C, Cesar HSJ, Pet JS. An economic analysis of blast fishing on Indonesian coral reefs. *Environmental Conservation* 2002, 26:83–93.

Pliny the Elder. *The Natural History*. Trans. Bostock J and Riley HT. London: Henry G. Bohn, 1842.

Pulmann P. *The Amber Spyglass*. London: Scholastic, 2001.

Pratt ML. *Imperial Eyes: Travel Writing and Transculturation*. London and New York: Routledge, 1992.

Ramage A, Craddock P. *King Croesus' Gold. Excavations at Sardis and the History of Gold Refining.* London: British Museum Press, 2000.

Read BE. Chinese materia medica, fish drugs. *Peking Natural History Bulletin*, 1939.

Richter C, Roa-Quiaoit H, Jantzen C, Al-Zibdah M, Kochzius M. Collapse of a new living species of giant clam in the Red Sea. *Current Biology* 2008, 18:1349–1354.

Riddle JM. *Dioscorides on pharmacy and medicine.* Austin: University of Texas Press, 1985.

Rose M, Acar O. Turkey's war on the illicit antiquities trade. *Archaeology* 1995, March/April:45–56.

Sainsbury K. The ecological basis of multispecies fisheries and management of a demersal fishery in tropical Australia. In *Fish Population Dynamics.* Ed. Gulland JA. New York: Wiley, 1998.

Salin KR, Yohannan TM. Fisheries and trade of seahorses, *Hippocampus* sp., in southern India. *Fisheries Management and Ecology* 2005, 12:269–273.

Samoilys MA, Martin-Smith KM, Giles BG, Cabrera B, Antica-mara JA, Brunio EO, Vincent ACJ. Effectiveness of five small Philippines' coral reef reserves for fish populations depends on site-specific factors, particularly enforcement history. *Biological Conservation* 2007, 136:584–601.

Sanders JG, Cribbs JE, Feinberg HG, Hulburd GC, Katz LS, Palumbi SR. The tip of the tail: Molecular identification of seahorses for sale in apothecary shops and curio stores in California. *Conservation Genetics* 2008, 9:65–71.

Shepard K. The fish-tailed monster in Greek and Etruscan art. Diss., Bryn Mawr College, 1940.

Smith WN. Origin of Seri Indian legends—The sea-horse story. *The Masterkey* 1958, 32:192–196.

Sowerby GB. *Popular History of the Aquarium of Marine and Fresh-water Animals and Plants.* London: Lovell Reeve, 1857.

Stott R. *Theaters of Glass: The Woman Who Brought the Sea to the City.* London: Short Books, 2003.

Tacon PSC, Wilson M, Chippindale C. Birth of the Rainbow Serpent in Arnhem Land rock art and oral history. *Archaeology in Oceania* 1996, 31:103–124.

Teske PR, Cherry MI, Matthee CA. The evolutionary history of seahorses (Syngnathidae: *Hippocampus*): Molecular data suggest a West Pacific origin and two invasions of the Atlantic Ocean. *Molecular Phylogenetics and Evolution* 2004, 30:273–286.

Teske PR, Hamilton H, Matthee CA, Barker NP. Signatures of seaway closures and founder dispersal in the phylogeny of a circumglobally distributed seahorse lineage. *BMC Evolutionary Biology* 2007, 7:138–157.

Teske PR, Hamilton H, Palsboll PJ, Choo CK, Gabr H, Lourie SA, Santos M, Sreepada RA, Cherry MI, Matthee CA. Molecular evidence for long-distance colonization in an Indo-Pacific seahorse lineage. *Marine Ecology Progress Series* 2005, 286:249–260.

Thwaite A. *Glimpses of the Wonderful: The Life of Philip Henry Gosse 1810–1888.* London: Faber and Faber, 2002.

Udo T. The seahorses: From the seahorse's mouth. *New Musical Express* (27 August 1996).

Unschuld PU. *Medicine in China: A History of Ideas.* Berkeley: University of California Press, 1985.

Uyarra MC, Coté IM. The quest for cryptic creatures: Impacts of species-focused recreational diving on corals. *Biological Conservation* 2007, 136:77–84.

Van Look KJW, Dzyuba B, Cliffe A, Koldewey H, Holt WV. Dimorphic sperm and the unlikely route to fertilisation in the yellow seahorse. *Journal of Experimental Biology* 2007, 210:432–437.

Vincent ACJ. Reproductive ecology of seahorses. Diss., University of Cambridge, 1990.

Vincent ACJ. *The International Trade in Seahorses.* Cambridge: TRAFFIC International, 1996.

Vincent ACJ. Trade in pegasid fishes (sea moths), primarily for traditional Chinese medicine. *Oryx* 1997, 31:199–208.

Vincent ACJ, Marsden AD, Sumaila UR. Possible contributions of globalization in creating and addressing seahorse conservation problems. In *Globalization: Effects on Fisheries Resources.* Eds. Taylor WW, Schecheter MG, Wolfson LG. Cambridge: Cambridge University Press, 2007. www.fisheries.ubc.ca/publications/working/series4.pdf.

Vincent ACJ, Meeuwig JJ, Pajaro MG, Perante NC. Characterizing a small-scale, data-poor, artisanal fishery: Seahorses in the central Philippines. *Fisheries Research* 2007, 86:207–215.

von Hippel FA, von Hippel W. Sex, drugs and animals parts: Will Viagra save threatened species? *Environmental Conservation* 2002, 29:277–281.

von Hippel FA. Solution to a conservation problem? *Science* 1998, 281:1805.

von Moltke K, Spaninks F. *Traditional Chinese Medicine and Species Endangerment: An Economic Research Agenda.*

London: International Institute for Environment and Development; Amsterdam: Institute for Environmental Studies, 2000. http://iied.org/pubs/pdfs/8140IIED.pdf.

Wang L, Lai C, Wu Q, Liu J, Zhou M, Ren Z, Sun D, Chen S, Xu A. Production and characterization of a novel antimicrobial peptide HKABF by *Pichia pastoris*. *Process Biochemistry* 2008, 43:1124–1131.

White J. *Journal of a Voyage to New South Wales*. London: J. Debrett, 1790.

White TH, ed. *The Book of Beasts: Being a Translation from a Latin Bestiary of the Twelfth Century*. Gloucester, UK: Alan Sutton, 1984.

Whitley G, Allan J. *The Sea-Horse and Its Relatives*. Melbourne: Georgian House, 1958.

Wilkinson C. *Status of Coral Reefs of the World: 2004*. Townsville: Australian Institute of Marine Science, 2004.

Yang S-Z. *The Divine Farmer's Materia Medica: A Translation of the Shen Nong Ben Cao Jing*. Boulder, CO: Blue Poppy Press, 1998.

Yellen JE, Brooks AS, Cornelissen E, Mehlman MJ, Stewart K. A middle stone age worked bone industry from Katanda, Upper Semliki Valley, Zaire. *Science* 1995, 268:553–556.

Zhang N, Xu B, Mou C, Yang W, Wei J, Lu L, Zhu J, Du J, Wu X, et al. Molecular profile of the unique species of traditional Chinese medicine, Chinese seahorse (*Hippocampus kuda* Bleeker). *FEBS Letters* 2003, 550:124–134.

Zhu C-H. *Clinical Handbook of Chinese Prepared Medicines*. Taos, NM: Paradigm Publishing, 1989.

INDEX

Printed in the United States
by Baker & Taylor Publisher Services